HOME EXTENSIONS

*Planning, managing and completing
your extension*

HOME EXTENSIONS

Planning, managing and completing your extension

Laurie Williamson

The Crowood Press

First published in 2000 by
The Crowood Press Ltd
Ramsbury, Marlborough
Wiltshire SN8 2HR

www.crowood.com

This impression 2003

British Library Cataloguing-in-Publication Data
A catalogue record for this book is available from the British Library.

ISBN 1 86126 291 4

All illustrations by the author.

Typeface used: Melior (*main text*); Helvetica (*boxed text and captions*).

Typeset and designed by
D & N Publishing
Membury Business Park, Lambourn Woodlands
Hungerford, Berkshire.

Printed and bound by Times Offset, Malaysia.

CONTENTS

About this Book 7
Introduction 8

CHAPTER 1: MAKING PLANS 9
Planning Ahead 10
Local Authority Approval 10
Planning Permission 13
Building Regulations 17
Structural Calculations 21
Building Plans 21
Hiring a Professional 22

CHAPTER 2: PREPARATIONS 24
Preparing a Flow Chart 24
Obtaining Quotations (Materials) 25
Obtaining Quotations (Tradesmen) 26
Value Added Tax (VAT) 27
Payments 27
Ordering Materials 28
Storing Materials 28
Room to Work 29
Plant and Equipment Hire 30
Safety and Insurance 31

CHAPTER 3: GETTING STARTED 32
Preparing the Site 32
Waste Disposal 34
Marking out Trenches 35
Excavation Works 37
Drainage Trenches 38
Soakaways 38
Safety 39

CHAPTER 4: DRAINAGE WORKS 41
How Drainage Works 41
Installing Drainage Pipes 43
Installing a WC 44
Connecting Two Drains 45
Installing a Back Inlet Gully 45
Manholes 46
Installing a Septic Tank 47
Rainwater Drainage 47

**CHAPTER 5: PHASE 1 –
BUILDING (FOUNDATIONS) 49**
A Gauge Rod 50
Concrete Foundations 51
Pouring Concrete 52
Ordering Concrete 53
Walls below Ground 53
Lintels 53
Underfloor Ventilation 54
Air Bricks 55
Preparing the Oversite 56
Concrete Floor Insulation 56
Damp Proof Course 57
Damp Proof Membrane 57
Laying the Oversite Concrete 58
Tradesmen 60

**CHAPTER 6: PHASE 2 –
BUILDING (CAVITY WALLS) 62**
Cavity Wall Construction 63
Mortar 63
Cavity Wall Insulation 64
Thermal Bridging 64
Vertical DPC 65
Wall Ties 66
Indents and Profiles 67
Lintels and Rolled Steel Joists (RSJs) 68
Door and Window Frames 70
Unglazed Window Frames 71
Window Templates 71
Door Frames 72
Scaffolding 72
Scaffold Hire 72
Wall Plate 75
Bricklaying 76
Tradesmen 76

**CHAPTER 7: PHASE 3 –
ROOFING 1 (FLAT ROOFS) 78**
Flat Roof Construction 78
Flat Roof Joists 78
Firring Section 80

Contents

Condensation	80
Cross Ventilation	80
Decking	81
Tilt Fillet	81
Felt Roofing	81
Cavity Trays	82
Lead Flashing	83
First Floor Joists	83
Herringbone Struts	83
Tradesmen	84

CHAPTER 8: PHASE 4 –
ROOFING 2 (PITCHED ROOFS) **86**
Pitched Roof Construction	86
Lean-to Roof	87
Close-Coupled Roofs	87
Hipped Roof Construction	87
Valleys	89
Gable Ladder	90
Fascia, Soffit and Barge Boards	91
Felt and Batten	92
Tiles	93
Slates	95
Cross Ventilation	95
Guttering	96
Carpentry	96
Quotations	97
Tradesmen	97

CHAPTER 9: PHASE 5 –
BUILDING (INTERNAL WORKS) **99**
Timber Floor Joists	100
Timber Flooring	100
Other Floors	101
Floor Insulation	101
Flat Roof Insulation	102
Staircases	102
Door Linings	104
Internal Load-Bearing Walls	105
Internal Non-Load-Bearing Walls	105
Tradesmen	106

CHAPTER 10: FIRST FIXING **108**
First Fix: Electrics	109
First Fix: Plumbing	110
Loft Hatch	111
Ceiling Boards	111

Fire Resistance	112
Loft Insulation	112
External Doors	113
Made-to-Measure Windows	113
Glazing	114
The Glazier	114
Double-Glazed Units	115
Tradesmen	115

CHAPTER 11: PHASE 6 –
BUILDING (PLASTERING) **117**
Window Boards	117
Studwork Walls	118
Plastering and Skimming	119
Artexing	120
Dry-Lining	121
Floor Screeding	122
Floor Tiling	123
Drying Out	123
Tradesmen	123

CHAPTER 12: SECOND FIXING **125**
Second Fix: Electrics	125
Second Fix: Plumbing	126
Internal Doors	128
Skirting Boards and Architraves	129
Dado Rails and Picture Rails	129
Handrails and Balusters	129
Wall Tiling	130
Painting and Decorating	131
Tidying Up	131
And Finally ...	132

APPENDIX 1: HOME
EXTENSION ACCOUNTS **133**

APPENDIX 2: FLOW CHART –
CHECK LIST **149**

USEFUL ADDRESSES **154**

FURTHER READING **156**

ADDRESS BOOK OF
TRADESMEN **157**

INDEX **158**

It is a fact that home extensions have always formed a good part of many local builder's and tradesman's workloads, but it was during the housing revolution of the 1970s that the home extension market started to grow at an alarming rate. It has continued to grow, through boom and bust, and now forms the backbone of new building works undertaken by the majority of small or medium-sized builders and tradesmen.

With such rapid growth inevitably came problems; poor work carried out by a small number of unskilled and unscrupulous contractors appeared with greater frequency, and at such a rate that court actions followed. These problems attracted a great deal of media interest and occurred with such frequency that the home owner, quite rightly, became more and more wary about employing unknown building firms to carry out what was likely to be the most expensive purchase they were ever likely to make.

To solve this problem the home owner should request references from previous customers to make sure that the builder is reliable. However, most opinions are normally based on three specific points: firstly, that the builder didn't over-charge, secondly, that the work was finished on time and thirdly, that the site was kept tidy.

Three important points, I agree, but with no bearing whatsoever on whether the builder or tradesman is, in fact, a good builder or a good tradesman. A good builder or tradesman must be judged on the quality of work done and not necessarily on how tidy he is. In the home owner's defence there has never been a step-by-step guide to help point out the important areas where the quality of the construction work is so vital.

Home Extensions is a step-by-step guide designed to assist anyone planning a new building project. It is filled with all the important points of 'good building practice' that must be observed during the construction works to ensure that the quality of the work is to an acceptable standard. *Home Extensions* is ideal for builders, sub-contractors and DIY enthusiasts, and is a handy, illustrated reference guide showing what is and is not accepted as 'good building practice'.

Many thousands of pounds can be saved by home owners supervising their own building projects, while employing skilled tradesmen to carry out work where their own personal skills may be lacking. Until now there has been no illustrated step-by-step guide to help them. Well, the wait is over.

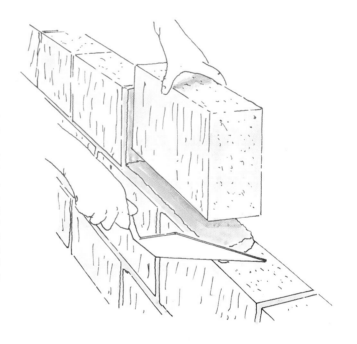

Introduction

Every year thousands of homes are extended and altered to accommodate an increase in family members, an improvement in living standards or perhaps just an investment in the property.

Whatever the reason for the planned extension work, we can now confidently undertake these larger projects thanks to the greater availability of information and assistance.

Couple this with the access we now have to reliable and trustworthy tradesmen, and projects which were previously out of reach for most do-it-yourself enthusiasts are becoming commonplace, continually producing first-class results at a greatly reduced cost. There are pitfalls, however, and these must not be underestimated.

The aim of *Home Extensions* is to provide a step-by-step illustrated guide to managing these projects successfully. The central theme is based on the use of experienced and qualified tradesmen where personal skills fall short. There are also clear areas where the keen DIY enthusiast can have a hands-on involvement.

MAKING PLANS

Building home extensions is a particularly British occupation, made easier by the use of a common national building design and the availability of the materials used. It has to be said that some freedom in the design of modern housing today has been sacrificed to satisfy the demands of the marketplace, resulting in the majority of houses being built around one central theme: a two-storey dwelling built in brickwork, blockwork or stone under a tiled or slated roof.

In European countries however, where large housing estate builders such as Wimpey and McAlpine are particularly uncommon, newly built houses are more likely to have been individually designed to maximize the use of each building plot, while still complying with similar building structure regulations and planning permission requirements as those specified on this side of the English Channel.

For the home owner this European model has a clear advantage over its British counterpart in that it will have been designed to meet the house purchaser's individual requirements, unlike the mass housing market in Britain where new home owners are unlikely to have had any say whatsoever in the design of the home they purchase. To balance this, the cost of house building in Britain is extremely competitive and compares very favourably with our continental neighbours. Quality is high and developers are constantly looking to meet the overall environmental demands placed upon them.

For the British home owner, building a home extension offers a definitive opportunity to make a personal mark on the property they have purchased, adding to, and rearranging the existing floor area to meet individual requirements or changes in lifestyle. Few of us live in our dream home, which is likely to be much larger than the house in which we actually live, so an extension will at least go some way towards satisfying this ideal without actually meeting all the special features many of us covet.

In reality the purpose of the extension may simply be to accommodate a growing

Fig 1.

family or an elderly relative, to increase room for leisure activities or maybe just to add space. Whatever the reason, it should be built to complement the existing property and will have to meet local building authority requirements.

The key to the success of this or any other building project will lie in the planning. To build a home extension employing a reputable builder would, in itself, place many demands and stresses upon any modern family and these will be greatly increased when the project is undertaken in a do-it-yourself manner. The building work, in fact, may be overseen by only one member of the family but it will assuredly affect everyone, so agreements and understandings should be reached and parameters set. Coping with massive upheavals can be managed where guidelines are drawn up and a framework timetable of the work is understood. It is therefore important to include everyone in the decision-making process, where the pitfalls can be explained along with the benefits.

> **DID YOU KNOW?**
> At the last count there were over 380,000 Listed buildings in England and Wales.

PLANNING AHEAD

Clearly the first, and most important points, in this decision-making process will be what the extension is for, where it will be built and what effect it will have on the appearance and value of the existing property. This will also be a good time to consider any longer term plans, for future extensions for example, so that these can be incorporated into both the current and the long-term picture. There are, of course, limits to all developments and home extensions must run the gauntlet of Local Authority appraisals before any building work can commence. Failure to get the necessary approvals can lead to heavy fines and even demolition of the unapproved building itself.

All home extensions have size limitations imposed upon them by the local authority, as well as by the space that is available for them. From this information you will be able to get a clearer picture of what is and is not likely to be permitted. If you are at all unsure and find it difficult to decide exactly what you want then, if you live on an estate where there are a number of similarly designed houses, it is likely that a variety of extensions will have already been built. It is unrealistic to assume that you can go looking for, and find, an extension which meets all your needs, but it is likely that a combination of certain aspects such as, perhaps, a window in one, a door in another, or the roof style of a third, will help you. Home extensions, as builders and planners will tell you, are not all the same, with many variations on a similar theme combining to form the picture you want.

If, as suggested earlier, you do live on an estate where several houses have been built around a general style, why not take the opportunity to wander round and see what other people have done. You may see something you like and, if you do find an extension similar to the one you are planning, ask the owners if they would mind showing you around to see how the new extension has altered their house layout. Ask questions about the new addition and how it has worked out for them. In the majority of cases, I am pretty certain, they will tell you that it is 'Just what we wanted!' and 'We cannot think how we managed without it!' And then there are the 'buts'. 'Wonderful ... but it would have been much lighter if we had gone for a slightly larger window.' 'Great ... but it would have been even better if we had put the door in the other corner.'

These are the points to remember. They may not be mistakes, as such, but they are possible minor improvements discovered as a result of other people's experiences which could help you when the time comes to finalize your own decisions. It is also important to discuss with them how the Local Authority reacted to their application and about any difficulties they encountered with the building work itself. 'Did the planners have any objections?' and 'How deep did you have to dig the foundations?'

It is usually only after the work is completed and the dust has settled that an objective view and the benefit of hindsight can be assessed. Of course if you live in an individually designed property, and not on an estate where easy comparisons can be made, decisions will be that much harder, but it is extremely likely that at least similar residences will be built nearby where improvements have already been completed and direct comparisons can be found.

During this decision-making process it is important that you do not lose sight of what effect, if any, the new building work will have on your neighbours. It may be that the extension structure in itself will not be a problem, but the work involved, the noise, the mess and the regular deliveries of bulky building materials could put a strain on even the best relationships, so try to involve neighbours in your plans and prepare them for any potential inconvenience to come. If planning permission is required for the extension your near neighbours will be consulted by the Local Authority, so early approval may be a good starting place.

LOCAL AUTHORITY APPROVAL

The range of home extensions for consideration is reasonably limited, with your choice likely to involve a single-storey or two-storey building with a flat roof or a pitched roof. Fortunately there are a great many design variations on these themes and as, in the majority of cases, your local Planning Department

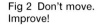

Fig 2 Don't move. Improve!

will have the final say on what you will and will not be permitted to build, this may well be the best place to discuss your proposals before any detailed plans are prepared. It is important to find out, as early as possible, whether your proposals are, or are not, in line with local development plans. The Local Development Plan will be available at your Planning Department and it will outline all aspects of the planning and environmental policies that the council has adopted and how they intend to carry them out.

The local Planning Department provides an invaluable service to the community and is charged with preventing excessive over-development, or what is now being called 'urban sprawl', while protecting Green Belt areas and retaining the balance and character of the area as a whole.

Of course not all home extensions require planning permission. Under local Permitted Development rules, extensions of a certain volume and in specific positions, rear extensions for example, can be

Fig 3 Will the extension be complementary and will it affect the value of your property?

Making Plans

Fig 4 Single-storey flat roof extension.

Fig 5 Single-storey pitched roof extension.

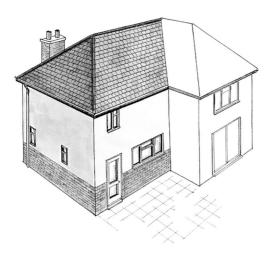

Fig 6 Two-storey hipped roof extension.

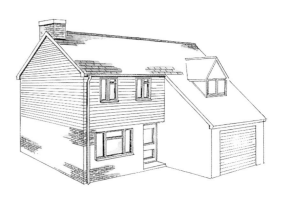

Fig 7 Two-storey garage extension.

Fig 8 Two-storey gable end extension.

Fig 9 Single-storey garage extension.

built requiring only Building Regulations approval. This also means that it is possible to build an unsympathetic and poorly designed extension which can seriously  affect the value of your property, create problems with your neighbours and alter the general appearance of the area as a whole. These occurrences are not necessarily unusual so, if you are at all unsure about how the new extension will affect your local area, and if you are looking for a professional opinion, it may be helpful to talk to your Planning Officer anyway. Apart from helping you with decisions, any meetings you have could help you when the time comes for the council to consider your planning application.

If you live in a Listed building or within a Conservation area, the Planning Department will automatically have a say in any development plans you may be considering because special rules apply in these particular cases. Armed with all this information you can now decide, with some degree of confidence, what size and style of extension is most appropriate for your needs, complements the existing  property, is fully in-keeping with the local surroundings while also, of course, being most appropriate for your pocket.

PLANNING PERMISSION

Where a planning application is required, the planning department will supply you with the necessary forms, tell you how many copies you will need to complete and return, and what the application fees will be. There are two types of planning application: Detailed and Outline. In most cases you will know exactly what you want and you will have discussed your proposals, maybe with the Planning Officer, to see what the council thinks of the building work you intend to carry out, so a Detailed application can be made. For a Detailed application you will  require a full set of drawings clearly showing the work you intend to carry out, and a plan of the site showing what buildings are already on the site and where the extension is going to be built. However, if you are not sure how the council will react to your proposals, and you want to gauge their reaction before going to the additional expense of preparing a full set of detailed drawings, an Outline Application can be made. If the outline application is successful a full set of drawings will be required and approval obtained before any work can commence.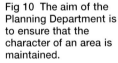

> **PLANNING PERMISSION**
>
> For a Planning Permission application you will be required to provide the following plans:
>
> 1. Existing elevation
> 2. Proposed elevation
> 3. Layout Plan
> 4. Block plan
> 5. Area Plan.

Making Plans

Under the Permitted Development scheme a large number of home extensions can be built without the requirement of planning permission, but check with your local council if you are not sure. The council may also remove Permitted Development rights where it feels the character of the area will be threatened by any new work.

It is likely that you will need to apply for planning permission if:

a. The extension roof is higher than the highest part of the original roof.
b. Any part of the extension is more than 4m high and is within 2m of the boundary of your property.
c. For a Terraced (including end of terrace) house where the volume of the 'original' house is to be increased by more than 10 per cent, or 50 cu m, whichever is the greatest.
d. For any other house where the volume of the 'original' house is increased by more than 15 per cent or 70 cu m, whichever is the greatest.

The 'original' house means the house as it was first built or as it stood at 1 July 1948. The increased development will *include* all other additions to the original house, even if they were built by previous owners. This calculation can include garages, sheds or any building which has been added to your property which is in excess of 10 cu m and where they are within 5m of the planned new extension. The volume is calculated by using external measurements.

You will also require planning permission if the additional building covers more than half the area of the land around the 'original' house. Special limits apply to Listed buildings and in Conservation areas, National parks and Areas of Outstanding Natural Beauty. *The following points are for guidance purposes only when you are considering outline ideas for your extension. Each Local Authority Planning Department will consider applications on merit alone.*

Extending a Detached House

Front extensions may be permitted but they will depend upon the space available and what effect, if any, they have

---

DOES MY NEW EXTENSION REQUIRE PLANNING PERMISSION?

These are just a few Planning Permission points to be considered when planning your new extension. These are points to indicate where planning permission will be required for your new home extension. They must only be considered as advisory points because in every case a planning application will be judged, by your local Planning Department, on its own individual merits.

1. If the new extension is within 20m of a highway, footpath, bridleway or byway.
2. If the new extension is over 4m high and within 2m of a neighbouring boundary.
3. If the new roof line is higher than the existing roof line.
4. If more than 50 per cent of the land, or garden, around the property is, or has been, developed.
5. Where the volume of the original house is to be, or has been, increased by 50 cu m for a terraced house or up to 115 cu m for any other type of house.

---

on the area. Side extensions should be of a size appropriate to the building and the plot and at least 1m from the boundary. Similarly, rear extensions should be of a size appropriate to the building and the plot and, in all cases, the extensions should be complementary to the existing building. Single-storey extensions may have either a flat or a pitched roof, but two-storey extensions will require pitched roofs with the roof line no higher than that of the existing building.

Extending a Semi-Detached House

Front extensions may be permitted but will be limited in size and location. Side extensions may be required to be set back from the front of the building and, along with rear extensions, will be limited in size depending upon the size of

---

OTHER PLANNING PERMISSION POINTERS:

1. Large front extensions are rarely acceptable.
2. New extensions should be compatible with the shape, style and size of the existing property.
3. The materials used should match or complement materials used on the existing property.
4. Every effort should be made to ensure that rear extensions do not 'overshadow' neighbouring properties.

Fig 11 Extensions should not overpower the existing building.

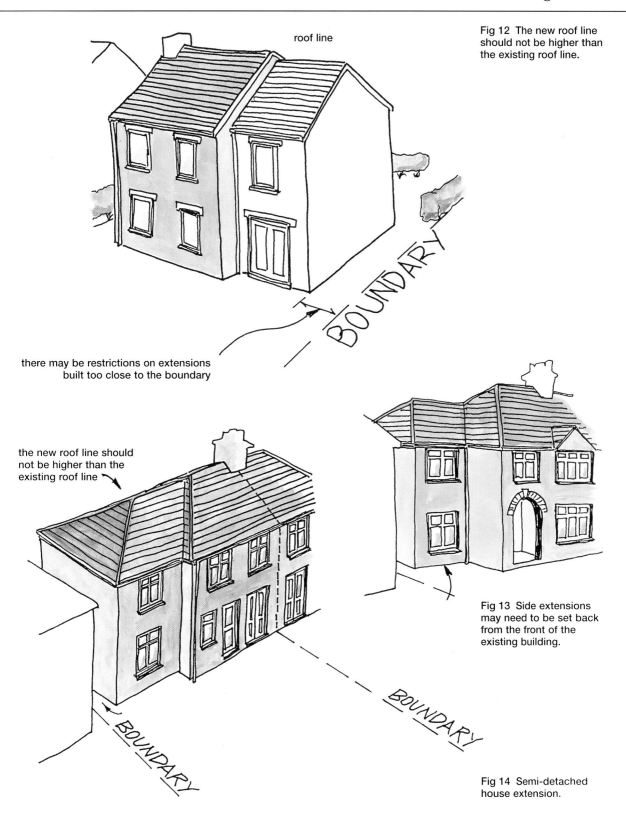

roof line

Fig 12 The new roof line should not be higher than the existing roof line.

there may be restrictions on extensions built too close to the boundary

BOUNDARY

the new roof line should not be higher than the existing roof line

BOUNDARY

BOUNDARY

Fig 13 Side extensions may need to be set back from the front of the existing building.

Fig 14 Semi-detached house extension.

15

the existing building and the plot. Side extensions may also be required to be at least 1m from the neighbouring boundary line. Rear extensions may also be limited by 'overshadowing' considerations.

Overshadowing

To protect a neighbours 'right of light', two-storey extensions should be designed not to overshadow windows, especially windows that have been there for over twenty years. Failure to comply could lead to legal action so check your rights before carrying out any work.

Extending a Terraced House

Front extensions are particularly uncommon and rear extensions will need to comply with stringent planning restrictions. It is extremely unlikely that an extension occupying the full width of the plot will be approved. The 'overshadowing' restrictions will also apply.

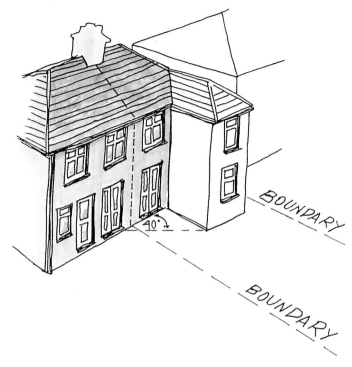

Fig 15 Take a neighbours 'right of light' into consideration when planning the size of your new extension.

try not to overshadow neighbouring windows

Fig 16 Rear extensions on terraced houses face extra restrictions on size.

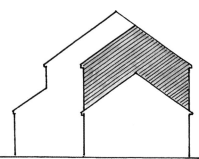

Figs 17 Building over a garage which shares a boundary with a neighbouring property may be permitted, but there are usually additional restrictions.

Building over a Garage

Where a garage shares a boundary with a neighbouring property, building over it may be permitted but there are usually additional restrictions.

When your plans are submitted the council should acknowledge receipt within a few days and then place the application on the Planning Register. This register is held at the local council offices and applications can be inspected by any interested member of the public. The council may also post notices about the proposed work close to the area where the work is to be carried out and even advertise in the local press. It will also notify neighbours of your intentions and request objections should there be any. The planning department will then prepare a report to be passed to one of the regular planning committees or it may, in straightforward cases, ask a senior planning officer to deal with the application. The council should decide your application within eight weeks, unless they request further time by written consent. The council or council officers will appraise proposals and consider whether there any good planning reasons for refusing permission. A refusal will be accompanied by reasons for the refusal and you may be asked to submit modified plans. If you feel that the reasons for the refusal are unsound you may appeal, within six months of the decision, to the Secretary of State. Up to a third of planning appeals are successful.

BUILDING REGULATIONS

Whether your new extension does or does not require planning permission, it will require Building Regulations approval. Building Regulations are rules approved by Parliament laid down to ensure that the minimum design and construction standards are achieved in domestic and commercial buildings. The Regulations are a list of requirements, referred to as Schedule 1, which are designed to ensure the health and safety of people in and around the building, including adequate access to and facilities for, disabled people, and provisions for energy conservation.

Also included on the list of works are: alterations to the drainage system; additions to washing and sanitary facilities; and construction work which alters the use of a building or is likely to have implications on adjacent properties, for example, work on party walls or underpinning.

TREES
Trees are an important feature in our landscapes and have a high amenity value in many areas. For this reason advice about felling or planting trees will be available from the Parks Department of your Local Authority. It is always advisable not to build too close to immature trees without first seeking expert advice.

Fig 18 Typical cavity wall construction.

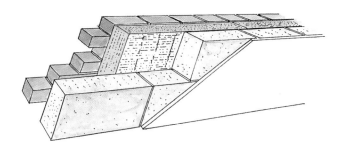

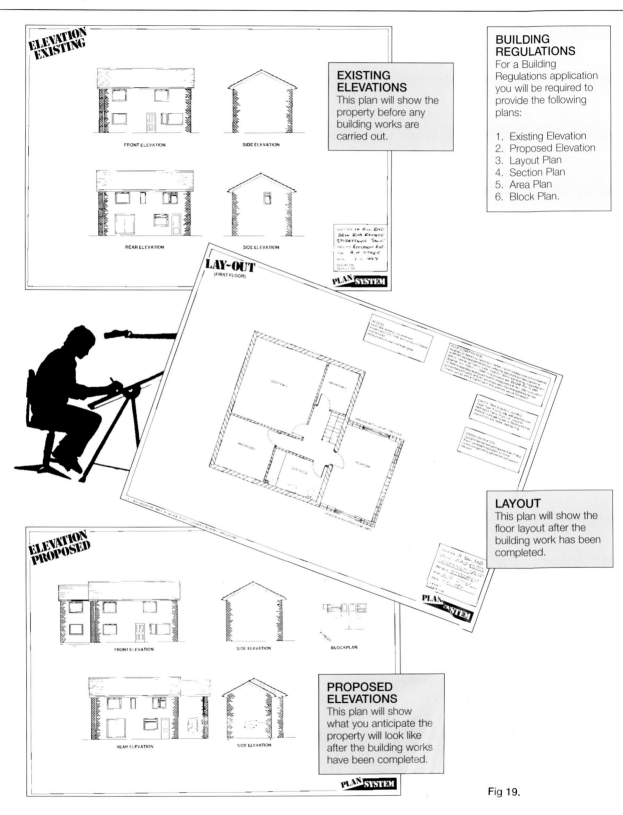

ELEVATION EXISTING

FRONT ELEVATION

SIDE ELEVATION

REAR ELEVATION

SIDE ELEVATION

EXISTING ELEVATIONS
This plan will show the property before any building works are carried out.

BUILDING REGULATIONS
For a Building Regulations application you will be required to provide the following plans:

1. Existing Elevation
2. Proposed Elevation
3. Layout Plan
4. Section Plan
5. Area Plan
6. Block Plan.

LAY-OUT
(FIRST FLOOR)

LAYOUT
This plan will show the floor layout after the building work has been completed.

ELEVATION PROPOSED

FRONT ELEVATION

SIDE ELEVATION

BLOCKPLAN

REAR ELEVATION

SIDE ELEVATION

PROPOSED ELEVATIONS
This plan will show what you anticipate the property will look like after the building works have been completed.

Fig 19.

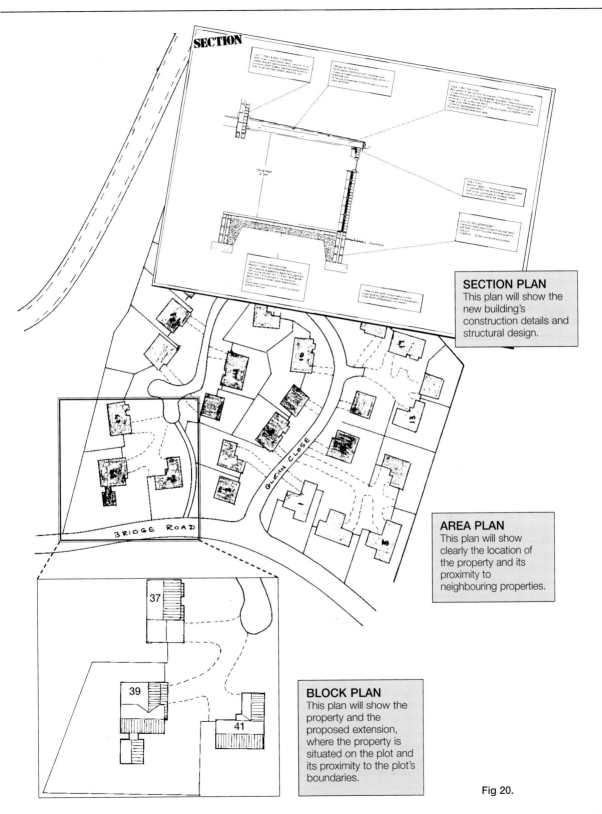

SECTION PLAN
This plan will show the new building's construction details and structural design.

AREA PLAN
This plan will show clearly the location of the property and its proximity to neighbouring properties.

BLOCK PLAN
This plan will show the property and the proposed extension, where the property is situated on the plot and its proximity to the plot's boundaries.

Fig 20.

19

Making Plans

Where a Building Regulations application is required there are three possible options for you to consider:

- The first, and most common for home extension works, is a Full Plans application. Detailed plans, clearly showing the proposed building works and all associated constructional details, will be submitted to the Local Authority Building Control Department for inspection. Building Control will be required to pass or fail the plans within a specified time limit and, if they find that the plans comply with building regulations, they will issue an approval notice. If, however, the plans are rejected they will issue reasons for the rejection.
- The second application is for a Building Notice where detailed plans are not required and no approval notice is given. One advantage of the building notice procedure is that you can carry out the work without the need to prepare full plans when you feel confident that the work to be carried out will comply with building regulations, for example, minor works.
- A third possible option is using an Approved Inspector. An Approved

Fig 21.

Inspector will check your plans, oversee the building work and, when it is completed to the satisfaction of the Approved Inspector, a final certificate of approval will be issued. If the work is not completed to the satisfaction of the Inspector a final certificate of approval will not be issued and the Inspector will then be obliged to inform the Local Authority so that they can consider their powers of enforcement.

Fig 22 Soakaway.

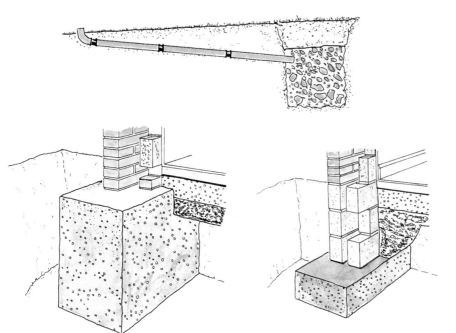

Fig 23 *(far left)* Trench fill foundations.

Fig 24 *(left)* Strip foundations.

Of course there are fees that will have to be paid to the Local Authority all relating to the type of work involved. A Full Plans application will involve two payments, a Plan fee and an Inspection fee. The first payment will be made when the plans are submitted for approval, while the second payment will be made following the first site inspection. Under a building notice only one fee will be payable, at the time you give notice, and will amount to the same as a Full Plans application fee. Details of prescribed fees, all subject to VAT, can be obtained from your Local Authority Building Control Department.

If you employ an Approved Inspector then the fee to be paid is a matter of arrangement between you and the Inspector and will also be subject to VAT. For home extensions where the use of sub-contractors and other tradesmen is planned and where detailed plans will be essential, a Full Plans application would be advisable.

STRUCTURAL CALCULATIONS

When the Building Inspector is not certain that the floor or roof timber sizes, walls or lintels specified on your plans are capable of carrying the weight imposed upon them by the new or altered structure, he may request structural calculations to satisfy any doubts. It is unlikely that these additional calculations will be included in the fee agreed with your architect or draughtsman, except by prior arrangement.

BUILDING PLANS

Planning Permission and Building Regulations applications will require detailed plans and, when you have a clear idea of what you want, these can be prepared. The plans required for a Planning Application are: an Existing Elevation Plan

Fig 25 Drawing your own building plans could save you money – if you know what you are doing.

showing the property before the proposed work is carried out; a Proposed Elevation Plan showing what the property will look like after the work is completed; a Layout Plan showing the proposed floor layout including the new extension; a Site or Block Plan showing where the new extension is to be built; and an Area Plan. The Area Plan is a copy of Ordnance Survey plans of the area and is generally available from your Local Authority.

Your Building Regulations application will require the same plans as those required for Planning Permission, plus a Section Plan. The Section Plan is probably the most technically detailed and represents a cross-section of the new work showing the important structural details. A list of the plans required for each application will be supplied by the local authority. All applications should be accompanied by the completed application forms plus the appropriate fees.

HIRING A PROFESSIONAL

To maximize the potential of your property, and to capitalize on the experience of qualified professionals, especially where personal experience is limited, it would be wise to use an architect or draughtsman to prepare your building plans. It is possible to draw them yourself, but the specific structural details required for Building Regulations alone can be daunting. By using a professional you can be confident that the additional expertise available will work in your favour.

Choosing between an architect and a draughtsman is not so clear and it is a decision often based on cost alone, with architect's fees generally higher than those of a draughtsman.

A qualified architect will have had extensive training and design experience and can be used to better advantage where these design skills are required. For small home extensions the draughtsman may prove to be the ideal and cost-

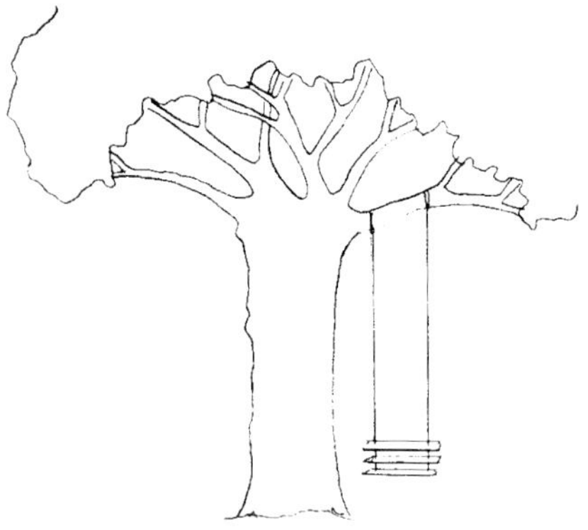

Fig 26 What the builder quoted for.

Fig 27 How the architect designed it.

Fig 28 How the sub-contractors built it.

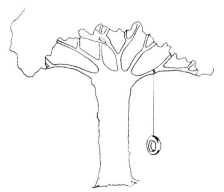

Fig 29 What the customer wanted.

effective solution, whereas extensions and alterations to older or Listed properties may benefit hugely from the design expertise of the architect.

Fees will be calculated according to the work involved so an accurate quotation of the fee must be obtained before you give the go ahead for plans to be prepared. Check what the fee includes and who will pay the Local Authority application fees, and ask if any provision has been made should structural calculations be required. An outline of the possible fee may be given over the telephone followed by a firm quotation after a site visit.

STEP-BY-STEP SUMMARY

Planning

Select the extension that best suits your home and is both practical and complementary. Look at the overall picture. Will this be the only extension or will you want to develop the property further?

Family

A significant part of any extension work will involve your family. Look at how the work will affect them and what benefits

there are to come. Also consider possible future additions to your family.

Property Value

How will the new extension affect the value of your property? Unusual or extra large extensions tend to detract from the original and you will not necessarily recoup the money spent when you sell.

Deeds and Covenants

Check with your deeds to make sure there are no covenants or planning restrictions relative to your property. The mortgage company should also be informed about your intentions.

Architect or Draughtsman

When you have gathered your ideas and plans together you will need to take the process one stage further. Select an architect or draughtsman and put your proposals forward for consideration.

Making Plans

Get your plans prepared and submitted for both Planning Permission and Building Regulations approval, if required. With these approvals you will be ready to move on to Step 2.

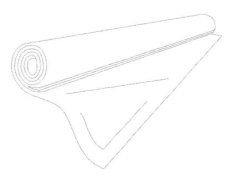

PREPARATIONS

It is at this early stage, after your plans have been drawn up and approval has been given by the Local Authority to carry out the building work, albeit within a limited time-frame (with Planning Permission when the work is not completed before a specific date a new application will have to be made) that preparations for the physical part of the building work should be made. This includes obtaining quotations from suppliers and tradesmen, ordering materials and, very important for all concerned, preparing a flow chart to calendar the work as it progresses. The accuracy of your flow chart will depend upon good planning, and the art of good planning is confidence. If you feel confident that this whole project is merely an opportunity to progress and succeed, rather than a problem to be solved, then the whole experience will be both beneficial and rewarding. (For complete examples of a flow chart, *see* Appendix 2.)

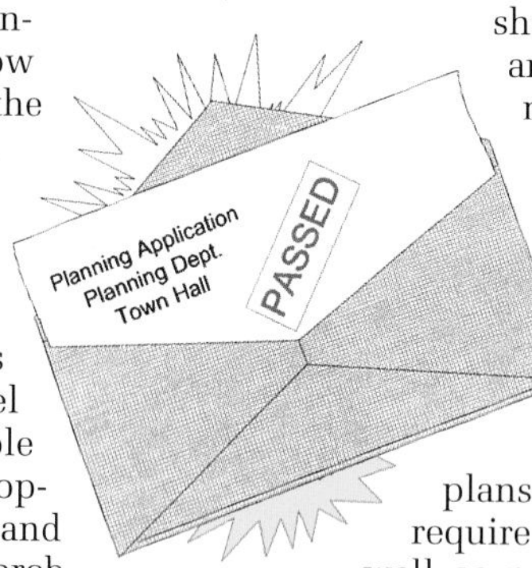

PREPARING A FLOW CHART

Firstly, and before you can add times and dates to your flow chart showing delivery of materials and arrival of tradesmen, the materials must be ordered from suppliers. With a plethora of suppliers and builders' merchants to approach, and a potentially large order to offer them, many of the larger suppliers will be happy to provide a 'quotation' using your approved plans for all the materials you will require to complete the project, as well as providing sensible and regular

Fig 30 'Now I've run out of flamin' blocks!'

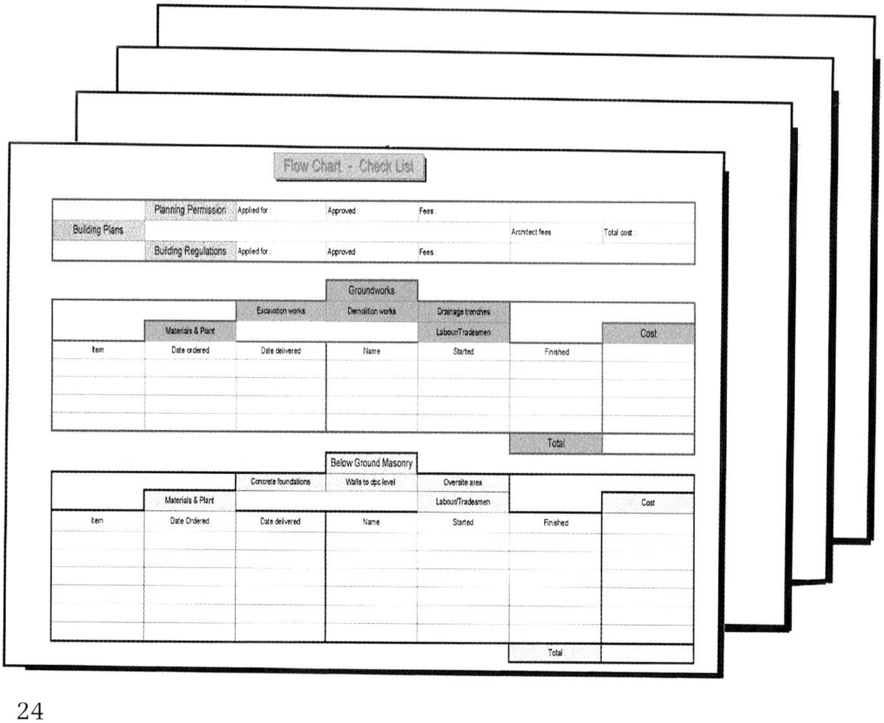

Fig 31 Flow chart.

delivery times to avoid overcrowding the site.

A small saving may be made by shopping around and breaking down deliveries from a variety of different suppliers, but this will only add to the demands and stresses upon you and could easily lead to an oversight resulting in a shortage of materials when the tradesmen are in mid-construction.

You may find that some materials will be in stock and some will need to be ordered. To avoid delays, and as an added precaution, when you take your plans to the supplier for pricing, take a sample of the bricks that you intend to use and, if possible, the roofing tiles; these are two significant areas where delays can occur due to stock shortages. Overall your joint aims should be the smooth running of the project and the avoidance of unnecessary delays. Take your plans to at least two suppliers and make sure delivery dates and quantities accompany the prices they give you.

As well as these standard material suppliers there are also specialist suppliers to be considered. Double-glazing units, for example, may require a special order from a specialist supplier. More so if the existing property is already double-glazed and you want to maintain the same style of units. The new unit sizes should be conveyed to the manufacturer or supplier and, if the units are to be installed after the bulk of the building work is completed, to avoid damage, template sizes will be needed so that templates can be made.

Other specialist suppliers such as the electrician, the plumber, the felt roofer and possibly the artexer, will, unless requested otherwise, supply and fix the materials they require. These supply-and-fix tradesmen make life a lot easier as additional orders are not needed, shortages will not arise, wastage is avoided and you only get charged for the materials used. Your supplier will know this and is therefore unlikely to include the items required for these trades in the quotation, unless requested to do so.

Secondly, and equally, if not more important, will be the recruitment of the tradesmen required for the formation of your 'team' to carry out this project. Compare this part of your flow chart preparation with a football manager preparing his 'team' sheet for a forthcoming football match and remember that your team will only be as strong as its weakest link and weak links can appear anywhere. A good tradesman will not want to follow on after a poor one. For this reason and for the project to succeed, the team members must be skilled and experienced at their particular trade leaving the emphasis on you, as the leader, to ensure that the project runs in a cohesive and balanced way and that, eventually, your aims are achieved.

With all this information and, following consultations with both suppliers and tradesmen, the whole project can now be laid out before you and times and dates added to your flow chart. Include some flexibility for unexpected delays and interruptions by the weather and, finally, remember that the times and dates are not cast in stone. If you have not built in enough spare time to cover the unexpected be prepared to bend a little and keep in regular contact with your 'team' to avoid misunderstandings.

OBTAINING QUOTATIONS (MATERIALS)

After your plans have been approved, select at least two major suppliers, builders' merchants, and give them each a copy of your plans, along with any supporting materials, such as bricks, roof tiles and so on, that may help them provide a detailed and accurate quotation. It is likely

Fig 32 Your local Builders' Merchant may supply a 'bulk price' quote for all the materials you require.

the quotes will include VAT at the standard rate and, for such a potentially large order, some discount should be negotiated.

Apart from the prices you will want to know delivery dates and when payments should be made. Variations from the quotation may result in additional charges so compare the quotes carefully to make sure all the materials specified match from one supplier to the other. One quote may be less than the other but you may find that an item has been overlooked and, in fact, the lower quote then becomes a higher one. Talking to, and by this I don't mean pestering, your suppliers is extremely important and deliveries should always be confirmed before the agreed delivery dates. Do not just wait and expect the delivery to arrive or wait until the day of delivery before checking, get confirmation of the delivery a day or two beforehand.

OBTAINING QUOTATIONS (TRADESMEN)

For the best possible results and, to ensure the smooth running of your project, use the best tradesmen you can find. Quality may cost just a little more but its value is timeless and will repay you time and time again in quiet satisfaction.

There is however, no easy way to select the best tradesmen other than by word of mouth and reputation. Before making your selection try to speak to a couple of previous customers and, if you feel it necessary, ask your local building control department if they can at least support your selection.

I do not think they are able to make recommendations as such, but I am sure they will steer you away from the unscrupulous operators if they can.

Make sure the tradesman views your plans and, if possible, visits the site. A site visit may not be necessary in the majority of cases as there will probably be very little to see but if, for example, there is a shortage of space for the bricklayer to work in he will be best advised, by you, to make a site visit to familiarize himself with any problems he may encounter when carrying out his trade.

The quotations must be given in writing and should clearly outline all the work included in the price. The first agreed method of payment is 'pricework' where a firm price is agreed for the work to be completed and that is what you will pay. The second method is a 'metreage rate' where the tradesman will be paid by the amount of metres or square metres involved in the project at an agreed rate per metre. The third, and more risky form of payment is called a 'day rate'. With this the tradesman will charge a specified amount for every completed day spent on site. This method of payment is more applicable to areas of work where a time cannot easily be determined accurately, groundworks for example. When paying a 'day

Fig 33 Keep in contact with suppliers to make sure deliveries arrive on time.

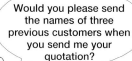
Would you please send the names of three previous customers when you send me your quotation?

rate' a close control of the work to be carried out must be maintained if you want to avoid the risk of large over-payments.

VALUE ADDED TAX (VAT)

If VAT is to be added to the price given make sure the quotation has a VAT registration number displayed on it and a proper VAT receipt is provided when payment is made. Some less worthy, non-VAT-registered operators, see the VAT system as a way to add more onto the bill which they then keep for themselves. A note of caution here.

PAYMENTS

The old adage that money greases palms is as true in the construction industry as it is in any other business. The important points, and it is my belief that more problems arise over money than any other area of the work, are that clear and succinct methods of payment must be agreed before deliveries are made and works commence. Suppliers will, naturally, want to be paid before they send you lorry-loads of expensive materials, but only pay for what is to be delivered and insist that an invoice is provided.

Paying tradesmen will require more care and making payments before work is carried out must be avoided. In special circumstances a tradesman may ask for a deposit, when ordering special materials for your particular project, but if this is not the case then any request for a deposit must be treated with extreme caution. There are two significant guides for payment to tradesmen and these, you will find, support standard practices.

The first, and most common with home extension projects, is when the tradesman expects to complete the work within one week, then one agreed payment in full will be expected and should be made available as soon as the work is completed to your satisfaction. The second method of payment, which is more applicable when the work is likely to take in excess of one week, will be staged payments where an agreed amount, consistent with the work that has already been completed, will be requested; this again is standard practice.

What is not standard practice and should be strongly advised against is payment for any labour 'before' the work you are being asked to pay for is completed. It is important to stress that not all building tradesmen are unscrupulous money grabbers who fail to turn up on time, overcharge and add unexpected extras to an agreed cost of the job.

There is practically no area of life where we can avoid ourselves being duped, but you can be sure that the vast majority of experienced and qualified tradesmen put the quality of their work before any payment and put 'trust' very high on their working agenda, while suffering from the over-exposure by the media, of the less trustworthy. Forewarned is forearmed and, although caution may be the byword, I am convinced that after the completion of the project you will number several new names among your list of friends.

> **STAGE PAYMENTS**
> Do not pay for work that has not been carried out. Stage payments should be an amount relevant to the work completed. Always retain a sensible final payment to be paid on completion.

Fig 34 'Yes Madam. You pay half before we start, then half when we finish!'

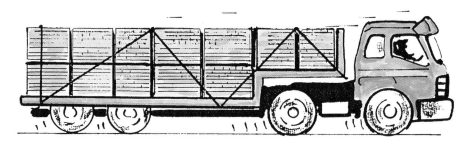

Fig 35 Deliveries must be supervised and materials checked for any damage in transit.

ORDERING MATERIALS

The art of managing incoming materials, and it is an art, may well determine the overall smooth running of your project. If space is plentiful and there is good storage for the materials that require protection from the weather then deliveries can be received several days or even weeks before they are required, although some of the more delicate materials, which include cement powder and plaster, are best fresh and delivered just prior to use. Bricks and blocks can be stored for quite a long time, albeit covered from very inclement weather, until they are needed.

The real art of managing materials is when storage space is at a premium. A common sight to us all is materials piled high in and around the front garden with heaps of sand and ballast close to, if not actually on, the pavement outside the house. In these instances and in particular where good, dry storage is limited, good planning is absolutely essential. It eliminates problems with neighbours, theft and damage to the materials themselves.

Where space is at a premium the deliveries will need to be well organized and storage, even when very limited, must be adequate to prevent not only weather damage but the likelihood of theft. The cost of materials stolen from building sites every year is staggering with the actual, unseen cost often being paid in delays to the building work itself. Try to ensure that materials are quickly stored away or covered up against the weather and where possible distributed around the site to where they will be required.

Fig 36 Double-glazed units should be purchased from specialist suppliers.

STORING MATERIALS

When considering the whole area of storage of materials, it is necessary to start with a plan which will include both undercover storage, for materials likely to suffer from weather exposure, and outside storage for materials which have at least some tolerance to poor weather conditions. In the majority of cases you will find that a garage or garden shed will provide the dry and secure protection needed with the overall aim being a steady rotation of materials as the building work progresses and new deliveries arrive. Keeping a close eye on each stage of the building work will ensure that the materials for that particular stage are handy and

Fig 37 Materials and sand heaps must be covered to avoid damage from inclement weather.

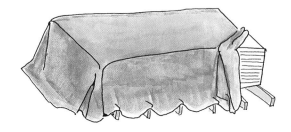

are not buried under materials to be used at a later stage of the construction.

The more bulky materials not likely to be stored inside, which includes blocks, bricks and sand, must be covered as a protection against the weather using tarpaulins or similar protective sheets; remember that hot, dry weather can be as damaging as cold wet weather. Particular care should be taken with sand, even when it is likely to be used within days of delivery, and especially in cases where it will be left overnight at the mercy of the local cat population. A protective sheet weighed down with blocks should provide adequate protection in this instance and can easily be removed when the sand is required.

ROOM TO WORK

It is likely, with the limited space available to the majority, that a balance must be reached between storage space and working space, but room for tradesmen to carry out their trades is primary and the greater the space available to them the happier they will be. An area around each site, and by site I mean the exact area where the new building is to stand, should be left completely free of hazards to accommodate standard building practices, which include the erection of scaffolding and transport of the various materials which will be used to construct the extension.

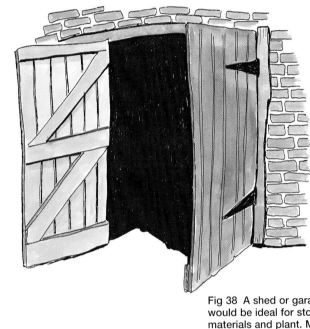

Fig 38 A shed or garage would be ideal for storing materials and plant. Make sure you keep it locked when not in use.

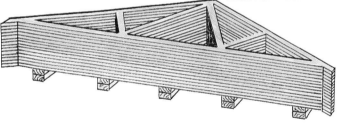

Fig 39 Stack trusses and all timbers on a level base, off the ground.

Fig 40 Cover sand heaps when not in use.

29

PLANT AND
EQUIPMENT HIRE

An absolute boon to the smaller building contractor and invaluable to the do-it-yourself enthusiast, plant hire ensures the availability and affordability of the larger items of plant and machinery we require to carry out areas of construction speedily and professionally. There are three significant methods of plant hire associated with the building and construction industry, of which two are commonly used for house extension works and the third, which is sometimes used for larger extensions, is more common with sizeable development projects.

The first is the hire of items to be used by you, or the tradesmen you employ, and includes cement mixers, tarpaulins, trestles and so on, while the second is items hired and erected by specialist firms including scaffolding and waste disposal bins. The third, and as I said earlier more common with larger development projects, is items hired to include an operative, or driver and includes larger plant such as JCBs (mechanical diggers) and tipper lorries.

Fortunately the availability to 'hire' the expensive plant and machinery you require to carry out smaller building projects ensures that the costs remain achievable and, therefore, brings larger building projects, such as home extensions, well into the do-it-yourself domain. It has to said that plant hire is now

PLANT HIRE CHARGES
Building plant such as cement mixers and angle grinders can be hired on a daily or weekly basis. To avoid excessive over-expenditure make sure plant hired is returned as soon as it is no longer required.

Fig 41 Tipper lorries can be hired for collecting bulky materials.

Fig 42 Make sure any plant hired is suitable for the purpose. Small cement mixers are ideal for bricklaying, but not for laying concrete foundations or oversite.

Fig 43 Larger excavations can be carried out using a mechanical digger.

so widespread and so very competitive that this can only be good news for us consumers. To add to our good fortune we can now hire the items we need on a daily or weekly basis and, with tight control, keep unnecessary expenditure to an absolute minimum. As a possible alternative to hiring the smaller items of plant such as drills and shovels, it may be a good idea to check the cost of 'purchase' with local suppliers, who are also becoming extremely competitive, and where good value is there to be found.

SAFETY AND INSURANCE

On-site safety is paramount both before and during the construction period, with the emphasis firmly on protection, for both the house occupants and the building operatives, should the unexpected occur. Public liability insurance for the entire period of the construction work is recommended and all safety equipment such as hard hats should be made available on-site for the duration of the work. These can be hired or bought from your local supplier, to be used when required, and all the operatives and tradesmen working on-site must be made aware that they are available. The tradesmen themselves are likely to have their own insurances and protective clothing but exceptional and unexpected circumstances do arise and these additional insurances will at least provide you with personal and financial protection.

STEP-BY-STEP SUMMARY

Passed?

When your building plans have been approved you will have a clear idea of the development work to come. Over the following period there will be a lot of information from tradesman and suppliers to be added to the Flow Chart.

Quotations

Shop around for quotations from both suppliers and tradesmen but do not stint on quality. The quotations should clearly state what is, and is not, included and tradesmen's references must be checked.

Delivery Dates

Before you finalize details with suppliers and arrange starting dates with tradesmen, check the availability of materials such as bricks, windows and doors. Build the delivery dates into your Flow Chart.

Storage

Protect all building materials against damage from the weather and from theft. Where storage space is at a premium good planning is essential. Materials stored outside should be covered with tarpaulins or similar sheets.

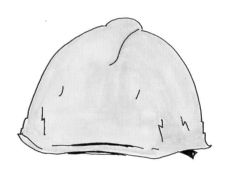

Fig 44 Safety helmets should always be available during the course of the building work.

Plant Hire

Hiring in plant such as cement mixers, scaffolding and trestles is a well practised method of getting large equipment on-site. If the excavations are extensive then maybe a mechanical digger and driver is a consideration. Whatever you choose, be sure to return equipment the moment it is no longer required.

Safety

You cannot take enough safety precautions on a building site. With heavy materials, holes in the ground, ladders and scaffolding the probability of accidents is high. Make sure you get public liability insurance, just in case.

> **SAFETY**
> Accidents are very difficult to avoid and, construction sites in particular, are very high on the list of accident black spots. To avoid adding to the already poor statistics, make sure the site is kept tidy at all times and every precaution is taken to warn visitors about the potential hazards. Hard hats should be available at all times and barriers should be erected around dangerous obstructions and open excavations.

GETTING STARTED

Having completed the building plans, received the necessary approvals, gathered together all the quotations and added the relevant information to your flow chart, it is now time to find out the difference between theory and practice. The success of each stage of the construction work will be the result of extensive forward planning and the first stage is no different.

When you consider the overall picture and what, eventually, you are trying to achieve, the affect the new extension will have on your garden, and this includes what affect your garden will have on the building work, must be high on your list of considerations. Many new, and by new I mean under ten years old, gardens will still be relatively immature and, for this reason, are unlikely to pose any serious problems on the site, such as hidden obstacles or relocation of well-established plants.

More care and attention to detail will be required where the property is older and the garden is more established. Small trees, shrubs and perennial plants will have had longer to become established and there may also be hidden problems involving old unused drains, wells and existing incoming services, such as electricity cables, gas and water pipes. It is also important that you should not treat your new extension completely in isolation because there will inevitably be some interaction between it and your garden.

To sum up briefly you should draw up a garden plan (a simple sketch plan will suffice), which should take all these points into consideration. It should also show the new positions for any plants and shrubs being relocated and any paths, patios or walkways either affected by, reinstated or newly formed as a direct result of the new construction works. It must be said also that, from a gardener's point of view, and for the best possible results, any plant relocation work will be dependent upon the prevailing weather conditions with good forward planning likely to produce the best results.

PREPARING THE SITE

The term 'site' has already been used and will be used again several times in this book so it is important to understand its meaning. First of all there is the site in general, which is the area allotted to carry out the whole of the construction works, and secondly there is the building site in particular, and this is the precise position where the new extension is to be built.

At this early stage the building site, in general, will have to be prepared, among other things, to cope with the safe and secure storage of incoming materials. Unnecessary obstacles should be removed and clearly defined areas must be allocated for deliveries. There should also be an area for rubbish and waste materials to be removed from the site and, importantly, an area for the construction work to be carried out. Some deliveries, premixed concrete for example, will only require enough room for access to the building site, in particular, without actually requiring room for longer-term storage, but the majority of deliveries, bricks and blocks for example, will need clearly designated areas for storage until they are required for use.

Delivery drivers will be limited to their areas of access and, even with crane-operated vehicles, they cannot be expected to search out the best areas to deposit materials. If you, or a representative, are not able to attend deliveries, make sure you leave sensible instructions that are well within the driver's dropping zone. When the materials to be delivered are vulnerable to damage by bad weather and under-cover storage is not possible, protective sheets or tarpaulins must be available to provide protection after off-loading.

At this stage the building site, in particular, must have all visible obstructions such as paving slabs, plants and shrubs removed so that the precise area for the excavation work can be clearly marked out. There may be, and it is extremely likely, hidden obstructions such as drainage pipes which must be

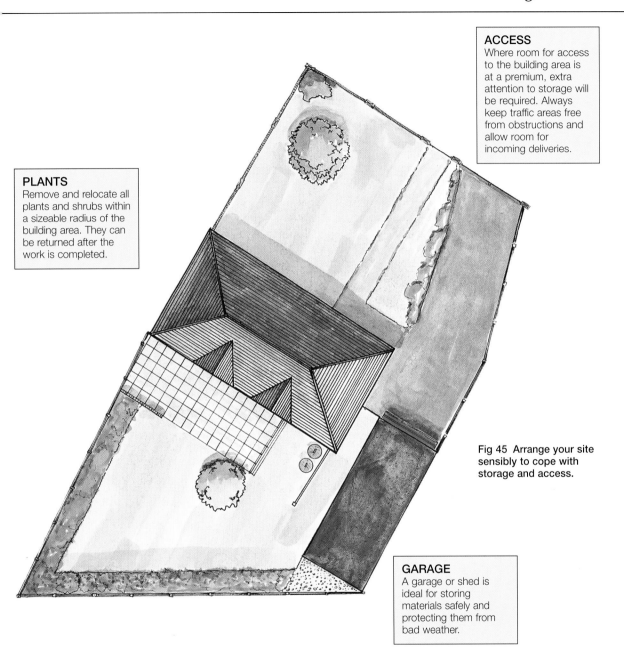

PLANTS
Remove and relocate all plants and shrubs within a sizeable radius of the building area. They can be returned after the work is completed.

Fig 45 Arrange your site sensibly to cope with storage and access.

GARAGE
A garage or shed is ideal for storing materials safely and protecting them from bad weather.

clearly marked as a reminder, to reduce the risk of damage during excavations. Using sand as a marker, and keeping to a neat tidy outline, mark out the whole area to be cleared and work only within this area.

Wholesale stripping and removing of plants and topsoil may produce more mess than necessary and result in extra work repairing the garden after the work is completed. The whole area to be built on must have all vegetation and topsoil removed to a depth that will prevent later growth and this includes materials, roots etc., likely to decompose and create problems with shrinkage as well as any compressible materials which could affect the stability of the building at a later stage. The topsoil can normally be relocated in selected areas around the garden.

SAND
When the builder's lines are in place, clearly showing the foundation positions, sprinkle sand along the lines, leaving an outline on the ground for excavations.

WASTE DISPOSAL

There is practically no area of the construction process where rubbish and waste materials will not be produced, often in abundance. A small proportion of the general building rubbish, such as broken bricks and blocks, can be used as ballast to assist drainage in the new soakaways or as a back fill to the foundation trenches. The top soil from the excavations can be dispersed around the site or can even be sold off to neighbours, but the majority of waste and rubbish, and this will include sub-soil from the trench and oversite excavations, should be removed from the site at regular intervals to reduce the risk of untidiness. Keep the building area tidy at all times and you will find that the tradesmen will continue in the same vein.

The surplus soil, and there will be a great deal from the excavations, will present the largest disposal problem and, unless you have a prearranged area for disposal of this waste, it should be removed from the site as it is excavated. For this purpose there are two widely used methods of disposal.

The first, and a very popular choice for smaller home extensions, is the use of waste disposal bins or skips. Skip hire companies are well known and plentiful and this method of waste disposal is extremely popular, providing an excellent and competitive service for this very purpose. The 'skips' can be hired for a set, and agreed, period of time and the hire charge will include any tipping charges, delivery to the site and collection.

One waste disposal bin can hold a large quantity of soil 'waste' and, with good timing and regular contact with the skip supplier, it can be replaced, often at very short notice, when it is full. Estimating how many skips will be required for the excavations may be tricky but a calculated guess should suffice. Soil increases in volume when dug so calculate the volume of the trenches and then multiply the total by about one-and-a-half times and this should give you an indication of the amount of soil to be removed and how many skips you are likely to require.

A word of caution if you decide to use skips. Try to arrange delivery of the skip early on the morning you expect the work to commence and then be replaced, if possible, without causing unnecessary delays to the excavations, early the following morning, then fill it as quickly as you can. The reason for this is that you should not leave the skip empty overnight or you may find that the neighbours have seized this opportunity to dispose of unwanted items such as old beds or old bikes, under the cover of darkness.

Another, and very important, point to consider for your skip delivery is location. Where will the skip be deposited? If there is room for it in the garden then this is the best place to put it but, if not, it may be necessary to leave it on the road outside. If this is your only option make sure you tell the supplier well before you expect delivery, and allow at least two weeks. He will then arrange for a licence from the Local Authority to leave the skip on the 'highway' and also arrange for lights, especially when overnight hire is expected. Do not leave a skip on the highway overnight without sufficient lights: accidents and prosecutions may follow. And, finally, if you find that it is not possible to fill the skip during the day, make sure it is covered over with a tarpaulin, or similar sheeting, overnight.

STREET PARKING
If you plan to leave the waste disposal skip on the highway you will need lights and a licence. The skip company will organize both if you give them enough notice.

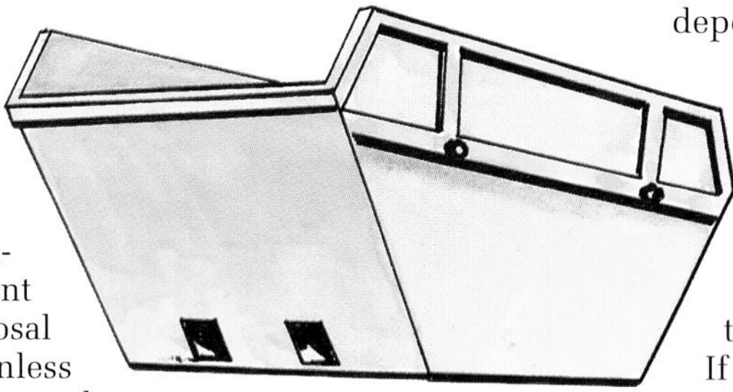

Fig 46 Waste disposal skips offer a competitive alternative when disposing of excavation waste.

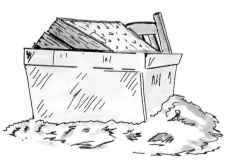

Fig 47 Leaving a skip empty overnight may be tempting for others.

The second choice is to hire a tipper lorry, with or without a driver. The waste soil and rubbish can be disposed of, for a tipping fee, at local tipping sites. Tipping in areas not designated for the purpose, called 'fly tipping' in the business and a major problem in some areas as tipping charges rise, can result in huge fines if you are caught.

MARKING OUT TRENCHES

With the site cleared of all obstructions and vegetative materials, including the topsoil, to comply with Building Regulations, the next thing to do is to mark out the trenches for excavation.

At this stage accuracy is very important if you are to ensure that the foundations are laid and the walls are built in the correct place. Check and double-check the trench dimensions and compare the positions with your approved plans, and remember that it is only after the concrete foundations are laid that

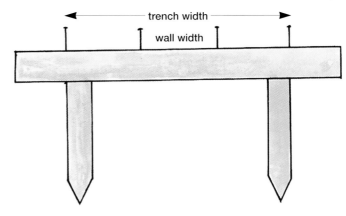

Fig 48 A profile board.

positional changes become both difficult and expensive. You will find the trench positions and dimensions clearly indicated on the approved plans.

At an earlier stage you will have roughly marked out with sand the whole area to be cleared for the extension, but this next part of the construction will require a greater degree of accuracy. To set the foundation trenches out properly you will require a good spirit level,

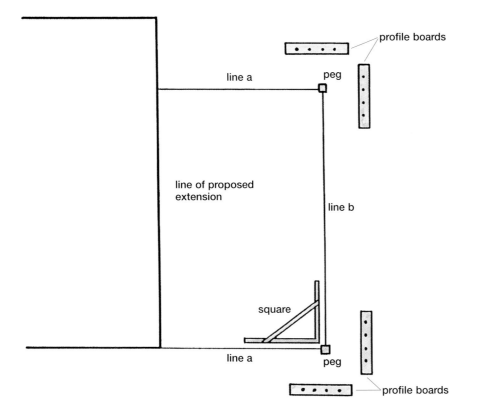

Fig 49 Double check all measurements when setting out the extension trenches.

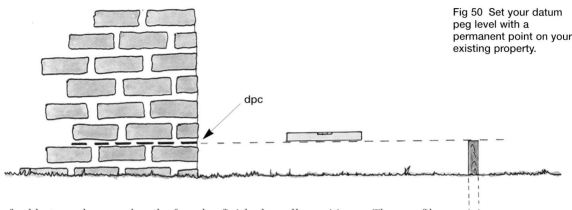

Fig 50 Set your datum peg level with a permanent point on your existing property.

dpc

preferably 1m or longer, a length of good building line, at least four profile boards and a datum peg. The profile boards will be used as guides for foundations and walls and the datum peg will be used to check levels.

The next step will be to select a level from which all future levels and measurements will be taken. As a rule, in the trade the existing damp proof course provides the best guide and the top of the datum peg will be set to reflect this level to other parts of the new building. It can then be used as a permanent guide from which all future levels are taken.

A 'datum' peg can easily be made from a length of wood, say 50mm × 50mm, hammered into the ground at a good depth for stability, while the top of the peg should be level with whichever existing guide you choose to use. Position your datum at the furthest point from the existing building but within a short distance of the area to be excavated. If you place it too far away it will be difficult to work from, however if it is too close there will be a risk that it may be knocked over or moved. All datums should be checked regularly in case of accidents, just to make sure that the new building level remains constant. You don't have to use the damp proof course as a guide, you may prefer to use the existing floor level, this is fine as long as the guide you choose is on the existing building and is a fixed point.

As soon as the datum peg is in position the profile boards can be erected. The top of each profile board must be level with the top of the datum peg. From these, and with the use of the builder's line, you can set out the foundation trenches and

finished wall positions. The profile boards, (see illustration), should be made from sturdy pieces of timber and, like the datum pegs, must be checked regularly for level and to ensure that they are square. For this, the most important part of the setting out, you will need a builder's square, (see Fig 51), to make sure that the corners are square and the new walls also abut the existing walls squarely.

Starting at the existing house wall, with the builder's square and plans to guide you, attach the first string line to the existing house wall where the new extension wall is to begin and at a point level with the datum peg, in this case the damp proof course level. Run the first string line (line a in Fig 49) away from the existing wall to a point at least 1.5m further than the extension measures.

At this point the first profile board can be erected, using the builder's square to ensure it is parallel with the existing house wall, and a nail can be hammered into the profile board to attach the line. Repeat this process for the opposite wall. The two string lines will now show two of the new extension wall

Fig 51 A builder's square.

5ft

4ft

3ft

A BUILDER'S SQUARE
To set the corners of your new extension accurately you can use a builder's square made out of three lengths of timber 3ft, 4ft and 5ft for the old, imperial measurements or 900mm, 1200mm and 1500mm for metric measurements.

positions. Measure along each line to the extension width and tie a small length of line as a guide. Using these ties as markers the side profile boards can be erected, again at about 1.5m from these string lines.

Running a third length of builder's string line to intercept the two markers to nails on the side profiles you now have the outer measurements of your extension (line b in Fig 49). Further nails and string lines can be added using measurements taken off the building plans. The two inner string lines show the internal and external position of walls and the outer string lines showing the footings. When all the lines are in position double-check the measurements with the building plans and square all the corners.

Next, remove the inner string lines for later use and, to mark the footing positions on the ground, trickle sand directly over the outer string lines. When a clear outline is marked these final string lines can be removed. You now have the footing positions marked ready for excavation. The builder's string line can be replaced at any time if you want to check the positions. Only after the concrete foundations are laid and the bricklayer has built the extension walls up to the level of the damp proof course can the profile boards be removed. Leave the datum peg in place until the extension is completed.

EXCAVATION WORKS

When all the preparations have been completed the excavations can begin. There are only a few areas of construction work where there is a choice of how you carry out the work and excavation is one of them. You can excavate by hand or by machine. Hand digging using shovels and a wheelbarrow can be extremely focusing and incredibly strenuous and is accompanied by a serious health warning. Recovering in hospital from a body breakdown can play havoc

Fig 52 'Your guvnor tells me you used to be a grave-digger.'

with Flow Charts so extreme caution is advised!

Digging footings by hand, or by any method, requires extreme caution. Careless excavations can cause damage to drainage pipes and incoming services. It is definitely not advisable to lacerate an electricity cable or a gas pipe with an over-energetic swing of a wayward pick-axe. For larger extensions, or just to speed things up, you may decide to excavate the foundations using a mechanical digger, or JCB. However, to do this, and I think it is reasonable to assume that the excavations will be carried out quickly, any time gained will be lost if the soil disposal is not carried out with equal speed.

For this reason, if you decide to use a mechanical digger, waste disposal skips are unlikely to be the best option, whereas a tipper lorry, hired with or without a driver, should prove ideal. A tipper lorry, and they come in various sizes, can remove quite a large quantity of waste soil to the local tip in fairly quick time, enabling the excavation works to keep moving on at a continuous rate. The tipper lorry is dual purpose in that, as well as removing waste materials from the site, it can also be used to bring materials such as sand and ballast to the site.

A word of caution. Before you decide that a mechanical digger is the best option, make sure that there is plenty of room on the site for the work to be carried out successfully, and that the digger, and the tipper lorry for that matter, have unobstructed access and plenty of room to manoeuvre.

DRAINAGE TRENCHES

Any new drainage trenches can be excavated at the same time as the footings, retaining the soil for covering the pipes after they are laid and checked. The trenches should only be wide enough for the pipes and deep enough, and at the required fall, for a 100mm layer of pea shingle on which the pipes will be laid. The trench depth will be determined by how deep the existing drainage pipes are, or according to the specifications on your plans.

Excavations for new Inspection Chambers (manholes) should also be made at the same time as the footing excavations, the concrete base for the IC can then be laid at the same time as the concrete foundations. At all times drainage excavations should be kept fairly tight as excessive room is not necessary.

SOAKAWAYS

There are really only two options available to you for the disposal of surface water from hard surfaces and rainwater drainage from roofs. The first is through the existing surface-water pipes and rainwater drainage system, while the second, and most common, is via a newly excavated soakaway. A soakaway, as shown on your plans, is a hole in the ground filled with rubble into which the rainwater and surface water will drain. That is the simple explanation and in the vast majority of cases that is exactly what happens. However, there is certainly more to it than that. The steering of surface water into a soakaway will only work as well as the permeability of the soil allows it to. Digging trial pits and filling them with water to test this theory is one, extremely technical, method, while the other method is to rely on local experience of soakaways, advice about which should be available from your local building control officer.

Each soakaway should be positioned at least 5m from buildings, unless specified otherwise, and be at least 1m

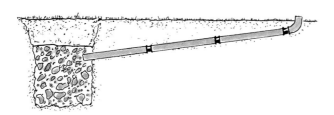

Fig 53 Soakaway.

cubed. Digging the soakaway before the building work begins is recommended as this will provide you with a hole in the ground into which broken blocks and bricks can be thrown. Make sure it is covered at all times for safety. The soakaway, when filled, must be covered with an old sheet of perhaps corrugated tin or similar, then covered over with at least 300mm of topsoil. All pipes carrying rainwater to the soakaway from the new building should be laid at a fall of at least 1 in 40 and must be suitable for the purpose.

SAFETY

Digging holes in the ground and excavating trenches provides the perfect opportunity for accidents to happen so great care is required. Any holes or trenches left unattended *must* be covered, or at least roped off, to prevent the unwary from stumbling into them. Greater care will be needed when weather or ground conditions fill the excavations with surface or ground water. Extensive barriers and warnings will not always prevent the unexpected when dangerous conditions prevail. Only leave holes and trenches unfilled for the shortest possible times.

STEP-BY-STEP SUMMARY

Site Clearing

If the area to be developed is part of your garden, there may be plants you will want to raise and set aside until after the

Fig 54 Keep all trenches and excavations guarded against mishap.

work is completed. Clear the site area of all obstructions and prepare it for setting out and excavation.

Waste Disposal

There will be a substantial amount of soil and rubbish you will wish to dispose of. Waste disposal skips represent a fast and easy way of getting rid of waste. Do not leave empty skips uncovered overnight or on the street without lights.

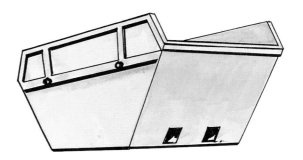

Setting Out

With the site area cleared you can set up the profile boards and line out the trench and wall positions with a string line. Using handfuls of sand you can then mark out the outline of the trenches before excavations begin.

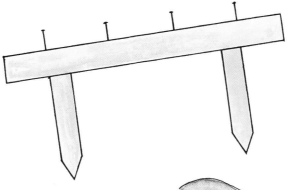

Dig Trenches

With the sand lines as a guide, the foundation trenches can be excavated. The topsoil can be used and the subsoil disposed of as you require. Be sure to let the Building Inspector know when you have started the excavation work.

Drains

The foundation and drainage trenches can be excavated at the same time, although you might find it easier to complete the majority of the drainage works before you start the building work. When the drains are laid and checked, fill in the trenches.

Rainwater

If your rainwater is being disposed of via a soakaway, it can be excavated at the same time as the other trenches. Any waste bricks and stones and non-perishable rubbish can also be thrown in.

DRAINAGE WORKS

Origination unknown, importance immeasurable. Probably the single most important factor in this country's health and well-being since the beginning of time must be the invention of our modern drainage system. It is not easy to find out who invented it, but progress from that day has continued and standards have steadily improved to a current situation where health and safety take priority.

To maintain this high standard, and to protect the general public, all new drains must be capable of conveying and discharging foul waste and surface water, without leakage or blockage, to sewers and drains in a manner which complies fully with Building Regulations. In addition, all the materials used must meet with British Standards requirements. Even small home improvements which do not normally needing Building Regulations approval, but where drainage work is involved, must comply fully.

HOW DRAINAGE WORKS

The discharge from your lavatory basin, hand basin, bath and kitchen sink is called Foul water and the discharge from gutters and gullies is called Surface water. The majority of this waste water

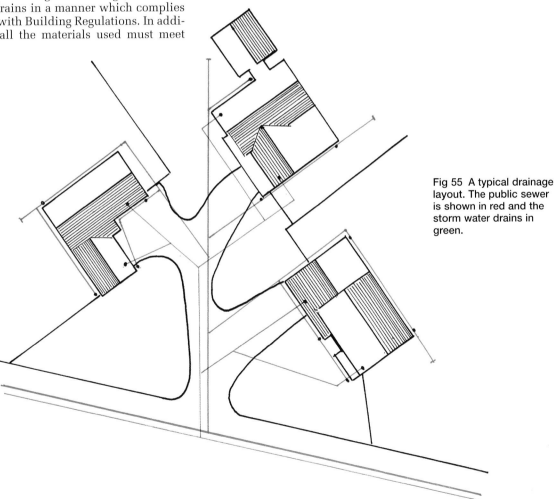

Fig 55 A typical drainage layout. The public sewer is shown in red and the storm water drains in green.

Drainage Works

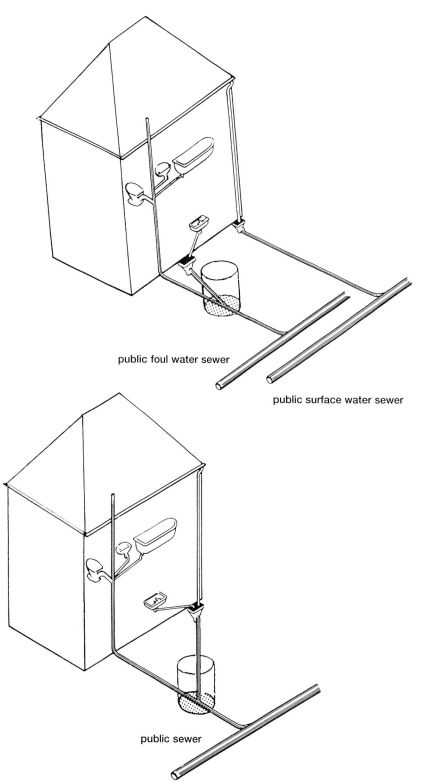

public foul water sewer

public surface water sewer

public sewer

Fig 56 Public surface water sewer.

Fig 57 Public sewer.

DID YOU KNOW?
In the region of £200 million worth of small building plant is stolen from building sites every year.

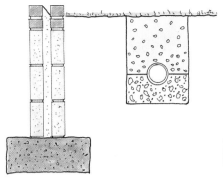

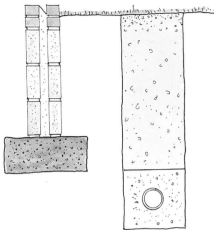

Fig 58 Where drainage pipes pass within 1m of, and are below the foundations, they will be encased in 1:3:6 mix concrete to a depth level with the base of the foundations.

is dispersed through Local Authority maintained sewers to sewage treatment plants. In our towns and cities, and urban areas in general, the method of dispersal by Local Authority sewers may vary, but there will be either a combined system for removing foul water and surface water waste through the same pipeline or there will be a system where separate sewers are provided for foul and surface water.

On many large housing estates the foul water may first discharge into a communal sewer before eventually being discharged into the public sewer. In rural areas where there are no public sewers, foul water will discharge into a cesspool or septic tank, to be removed by approved contractors, and surface water will discharge into a soakaway.

Any additions to your existing drainage system will require Building Regulations approval before any work can start. The design and layout of these additions should be kept as simple as possible, with short pipe runs and plenty of access to deal with blockages, should they occur.

INSTALLING DRAINAGE PIPES

New drains carrying foul waste away from buildings to the sewers must be installed effectively and without leakage, which is why drain laying is a skilled job. The specification of your new drains will be clearly shown on your approved building plans showing a choice of materials, either vitrified clay pipes or upvc are the most common.

It is preferable, but not always possible, for drains to be laid outside the building structure, but where pipes pass under buildings and through walls, adequate protection must be afforded to them. The pipes must be laid on a granular bed, pea shingle for example, at a

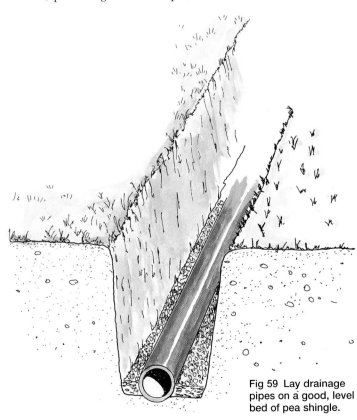

Fig 59 Lay drainage pipes on a good, level bed of pea shingle.

Drainage Works

fall of not less than 1 in 40 and, where drainage pipes are laid within 300mm of the ground level, in areas likely to be affected by human or motorized traffic, they must be covered with concrete for protection.

To avoid de-stabilizing existing and new foundations, drains passing within 1m of the foundations and below the foundation level must be encased in concrete up to, and at least level with, the base of the foundations (*see* Fig 58). Where new drainage pipes pass through walls a lintel will be installed above the opening with a gap of at least 50mm around the pipe to allow for any settlement (*see* Fig 70). A rigid sheet can be fitted around the pipe covering the open gap to prevent access by burrowing animals.

A major part of the installation process is the trench digging. By allowing about 100mm extra trench depth for the addition of a granular bedding base for the new pipes to be laid on, and 150mm around the pipe in order to provide sufficient space for jointing and proper compaction of bedding materials, the result should be an even and secure system. The gradients, as specified on your building plans, can be prepared by using a gradient board or a string line to confirm that the pipes are laid at the correct 'fall'. Add the granular bedding, pea shingle is commonly used, and then lay the pipes as carefully as possible to avoid breaks and cracks. The Building Inspector will inspect your new drains and may carry out a drain test on them to ensure that the seals are secure and the risk of leakage is minimal.

INSTALLING A WC

Many new home extensions include the addition of a WC or Water Closet and you will find that Building Regulations are

Fig 60 *(Below left)* Installing a new drain for a WC.

Fig 61 *(Below)* Installing a new WC and inspection chamber.

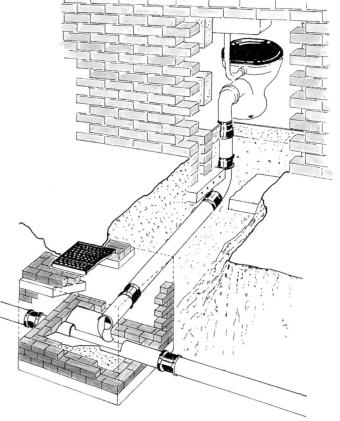

particularly strict on two points. Firstly, the location of the 'smallest room' in comparison with the other 'habitable' rooms and, secondly, the source of ventilation. A toilet must not open directly into a habitable room such as a kitchen or lounge and windows must be of a style and size to provide adequate ventilation. In the absence of mechanical windows, electric ventilation will need to be installed to provide the number of air changes required, per hour.

The lavatory basin will have either a 'P' trap or an 'S' trap. The 'trap' is very important because it provides a water barrier between the internal air and the drains, preventing noxious gases and other nasty smells entering the toilet area. The only difference between a 'P' trap and an 'S' trap, as far as installation is concerned, is the type of connector required to join the 'pan' to the drains.

CONNECTING TWO DRAINS

New home extension drains are invariably small additions to the existing drainage system requiring uncomplicated connections where blockages are unlikely to occur. With such simple drainage additions you may only require a drainage junction such as a 45 degree connector. The important factors to consider in these cases are: what type of waste is being carried through the drain, and what precautions can be taken to avoid a blockage, should one occur?

Building inspection chambers at all drainage junctions, and all over the site to cope with every eventuality, should not be necessary even where junctions do occur, and junctions are probably the major cause of blockage problems. To deal with these less risky cases the installation of a Rodding Eye, a pre-formed section made specifically to provide access for the rodding of drainage pipes, can be installed and will provide access to the drains to deal with blockages.

INSTALLING A BACK INLET GULLY

A Back Inlet Gully is installed when a kitchen sink, bath or wash hand basin discharges directly into the drainage

TRAPS
When you select a new lavatory basin it will have either a 'P' trap or an 'S' trap. The trap is the section of the bowl where a water barrier prevents noxious odours and gases entering the toilet area from the drains. Make sure your plumber is aware which trap style you require or have ordered.

Fig 63 A rodding eye.

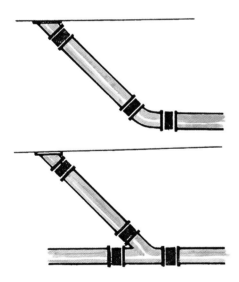

Fig 62 A rodding point at the end of a drain run (top) and along the drain run (bottom).

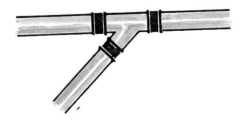

Fig 64 Connecting to a drain run by adding a junction.

Fig 65 A back inlet gully (BIG).

Drainage Works

system and not through the soil vent pipe. The BIG has a trap at its base similar to the trap of a toilet and, when filled with water, acts as a barrier preventing gases and smells entering the building from the drainage system. In some cases, kitchen extensions in particular, it may be necessary to leave an existing back inlet gully *in situ* and for it to remain in use. When this is the case the BIG will be raised to floor level and fitted with a secure, screw-down lid in place of the standard drain grid. The screw-down lid will provide access, should it be required, to the gully and surrounding drains should blockages occur.

MANHOLES

An Inspection Chamber, more commonly known as a Manhole, serves two purposes. Firstly, it provides a junction where several drains will meet and form into one drain, and secondly it provides a point of access to the drains where rodding canes can be inserted to deal with blockages. Where inspection chambers are less than 1m deep there is a choice between pre-formed polypropylene units and brick-built units; when the chambers exceed 1m in depth brick-built or pre-formed concrete chambers should be used.

Built on-site off a concrete foundation, the brick-built inspection cham-

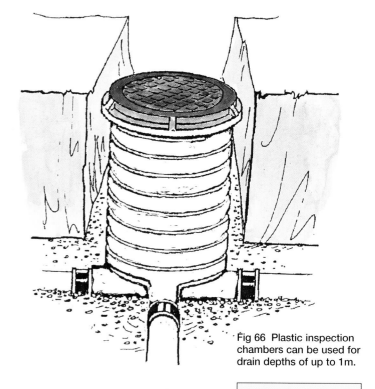

Fig 66 Plastic inspection chambers can be used for drain depths of up to 1m.

PLASTIC INSPECTION CHAMBERS

Plastic inspection chambers offer a good alternative to brick inspection chambers. They are cheaper by comparison, lightweight and easy to install, but there is a limit to the depth at which they are permitted to be used.

Fig 68 Plastic inspection chambers can be easily adjusted to meet site requirements.

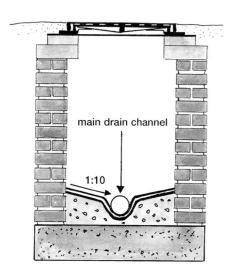

Fig 67 A brick-built inspection chamber.

bers will be built approximately 225mm thick in English bond-style brickwork using semi-engineering or similar bricks which are suitable for this purpose. In the base of the inspection chamber the channels open out and discharge their waste into a central main channel. Standard household drains are generally 100mm in diameter, although larger sizes are available where specified. Side channels connecting the main channel will be installed to direct their waste with the flow of the drain and, with foul waste in particular, the channel junctions should not exceed 90 degrees.

INSTALLING A SEPTIC TANK

In rural areas where there are no Local Authority sewers, properties deal with their own household water waste using a cesspool or a septic tank. The old fashioned cesspool was a brick or block-built chamber set below ground level into which foul water waste would discharge. The solid waste would settle in the bottom of the 'pool' and the water drained off into the surrounding area by way of land drains. The modern cesspool and septic tank are now environmentally friendly units made from fibreglass and designed specifically for the purpose. The septic tank performs in the same way as the old cesspool and, when necessary, is emptied by approved contractors. The contents are then disposed of or spread over farmland as fertilizer.

RAINWATER DRAINAGE

Rainwater discharged from roofs and other areas will be diverted into surface water drains and then into the public sewers or into a soakaway. When a soakaway is required it must be positioned to ensure that the disposal of the surface water and rainwater will not affect the integrity or stability of any foundations or buildings within a 5m radius of the soakaway. If you have any doubt about the permeability of the ground and its ability to cope with this additional water your Building Inspector may be able to advise about the operation of soakaways in your area.

STEP-BY-STEP SUMMARY

Manholes

Build a new inspection chamber off a proper concrete foundation. Drains should be stable and unlikely to suffer from subsidence. New branches built into

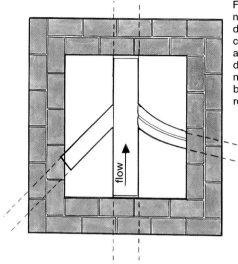

Fig 69 When adding a new branch to an existing drain in an inspection chamber and the angle of an existing branch is 45 degrees or more, then the new branch channel must be of three-quarters round section.

existing inspection chambers must meet Building Regulation requirements.

Blockages

Make sure there is access to all drains. Rodding eyes can be installed and are essential where access is likely to be difficult. Blockages usually occur where the waste changes course in the drain run.

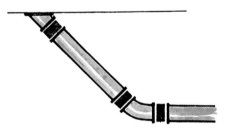

Drain Test

Make sure new drains are watertight and fall at the correct gradient. A gradient board can be made to check this. The Building Inspector may wish to test this to make sure there is no likelihood of localized soil contamination occurring.

Under Buildings

Drains under buildings must be protected by concrete reinforced lintels. Allow a small gap around the pipes for settlement and shield this area to prevent burrowing animals and insects getting under the building.

Drainage Works

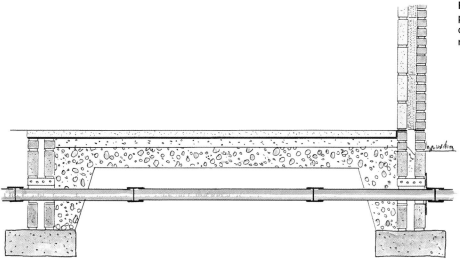

Fig 70 Where drainage pipes pass through walls concrete lintels will be required.

New WC

When installing drains for a new toilet, make sure the connections match with the *trap* of the lavatory basin. A 'P' trap or an 'S' trap.

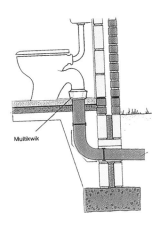

Multikwik

Cover

Manhole covers situated in areas where there is traffic, whether motorized or foot traffic, must be suitable for the purpose. Manhole covers inside buildings will be double-sealed to prevent noxious gases and waste entering the building should blockages occur.

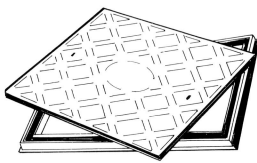

Soakaways

Always keep open holes, soakaways for example, covered until they are filled. When the soakaway rubble is in place it will be covered with a corrugated tin sheet or similar and then covered with topsoil. Mark the soakaway location on your plans for future reference.

PHASE 1 – BUILDING (FOUNDATIONS)

As soon as you have completed the excavation works the trenches must be inspected and approved by your Building Inspector. You will be asked to provide approximately 24 hours notice to Building Control for the inspection to be carried out before the foundation concrete can be laid. If the trench base is not secure and has not been excavated to an acceptable standard to ensure the integrity and stability of the foundation concrete, you will be asked to make whatever improvements are required and this may include digging the footings a little deeper until the base is secure.

Then, and only then, should you make plans for pouring in the concrete foundations. At this time the builder's lines

Fig 71 Always build off solid foundations.

FOUNDATION DEPTH?
The depth of the foundations may be specified on your building plans. However, the Building Inspector will have the right to alter these specifications to make them compatible with site conditions.

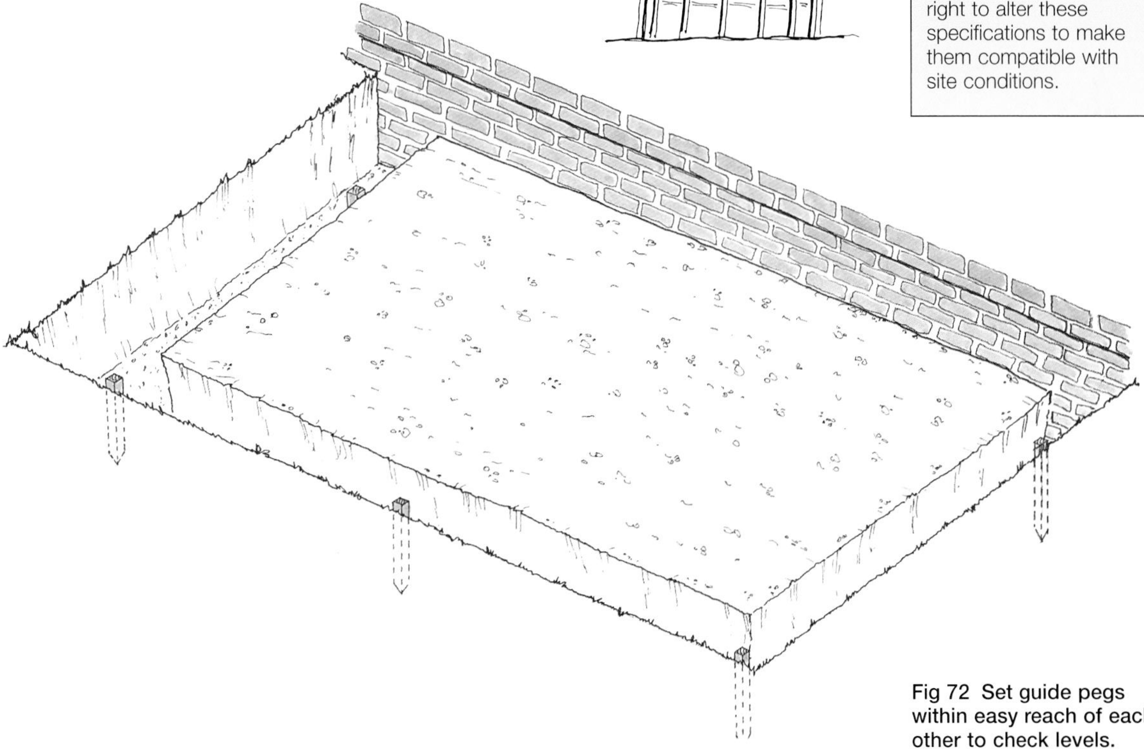

Fig 72 Set guide pegs within easy reach of each other to check levels.

Phase 1 – Building (Foundations)

you removed from the profile boards to carry out the excavation work can be reinstated, so that the depth of your trenches can be measured and the amount of concrete you require can be calculated. Stakes can be driven into the base of the trench at sensible intervals of about 2m, to provide a guide for the concrete foundation depth and level. To calculate the height, each stake must be set at a measure down from the reinstated builder's line then the stake driven firmly into the trench base. When the first stake is set at the required level use a straight edge and a spirit level to drive additional stakes into the trench base until the whole foundation area is set out.

Fig 73 To remove the use of split courses set the guide pegs to suit a minimum depth of foundation while coinciding with brick and block measurements.

A GAUGE ROD

To assist you with the setting up of foundation depths and wall heights the best tool to use is a gauge rod, made out of a piece of spare timber. Using this, you can plan the wall height incorporating the materials to be used. It will be helpful to the bricklayers if you can calculate the foundation depth to coincide with brick or block measurements. For example, if the walls to be built in the footings (the

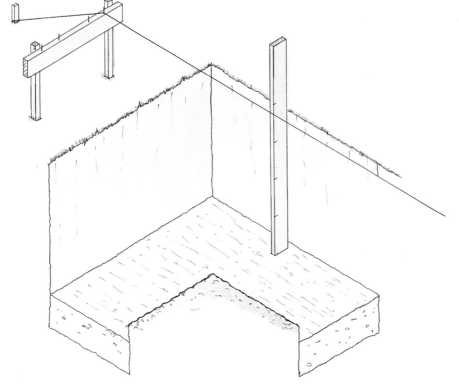

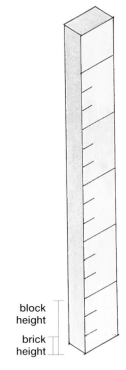

block height

brick height

Fig 74 Make a gauge rod to calculate concrete foundation depths.

Fig 75 Measuring down from your dpc line you can calculate the number of block and brick courses required.

50

walls below ground level) are concrete blocks with a brick finish where the wall is visible, your gauge rod can be made showing the number of blocks to be used, plus the number of bricks to be used, plus the cement mortar bedding between each course (*see* Fig 75).

You may find that increasing the depth of the foundations will avoid the necessity for split courses. A split course of bricks may be used where the courses are calculated to be less than the height of a brick. This does happen occasionally, is permissible below ground level, and should not affect the overall integrity of the wall. By setting the trench stakes to a depth acceptable to the measurements on the gauge rod the courses of blocks and bricks should make up to complete courses.

CONCRETE FOUNDATIONS

As far back as the 1920s concrete began to replace brick, and in some cases stone, as a foundation for walls and buildings. Previously hand-mixed and now more commonly mixed in a concrete mixer, the foundations are there to bear and spread the load of the wall built off it. A typical concrete mix for foundations is known in the trade as a 1:3:6 mix and consists of one part Portland cement (Portland cement is a complex, heat-treated mixture of lime, silicates and aluminium with iron oxides added) and nine parts ballast (already mixed at three parts sand and six parts gravel). When mixed with water the mixture undergoes a fairly rapid chemical reaction which transforms it into a hard, rock-like substance which has great strength.

Fig 76 Make sure any plant hired is suitable for the purpose. Small cement mixers are ideal for bricklaying, but not for laying concrete foundations or oversite.

Home extension foundations can be mixed on-site or delivered, ready-mixed, to be poured into the foundation trenches to a depth required by the Building Inspector. A specified foundation depth may be shown on your building plans, but this is flexible in that you can increase the thickness but not decrease it.

There are two common types of concrete foundation, each having benefits of their own. The first, and most popular, is a strip foundation. Set in the bottom of the trench at a minimum, agreed, depth as specified on your building plans, the walls will be built centrally off the foundation to ensure stability. The second type is a trench fill foundation which is, as it sounds, a concrete foundation filling the trench to within a short distance of the top. It is possible to build off the edge, rather than centrally, of a trench fill foundation and is commonly used where walls are built as close to a boundary as is possible. The

DID YOU KNOW?
Over 80,000 tons of lead are used in the UK building industry every year.

Fig 77 Order pre-mixed concrete for foundations and oversite.

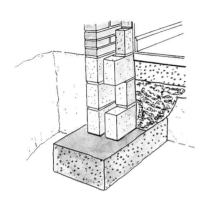

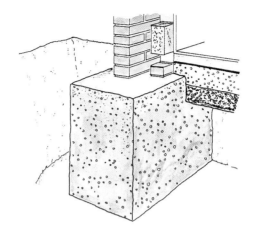

Fig 78 *(far left)* Strip foundations.

Fig 79 *(left)* Trench fill foundations.

drawback with a trench fill foundation is where drainage pipes pass through the foundation. When this is the case the pipes should be well wrapped in a protective material padding and positioned when the concrete is poured, to avoid any pipe damage due to settlement of the foundations.

A third and less common foundation is a Raft foundation. Raft foundations are more likely where landfill and subsidence are possible and consist of a reinforced slab up to 300mm below ground level, thickened up where the walls are to be built. A polythene membrane is laid on a sand blinded hardcore base then the first concrete layer of the raft is poured. A layer of reinforced mesh is set in this layer then another layer of concrete is poured. A second layer of mesh is set into this layer then a final top layer of concrete is poured on to finish.

POURING CONCRETE

Speed is of the essence when laying or pouring concrete. Firstly, mixing and then pouring the concrete should be a continuous process to ensure that each load bleeds in well with the previous load, maintaining the strength of the concrete. Do not pour in a load then leave it for an hour or so then pour in the next load. Concrete soon becomes dry and unworkable, forming a join where the two loads meet which will become a weak point in the foundation. When you are mixing the concrete by

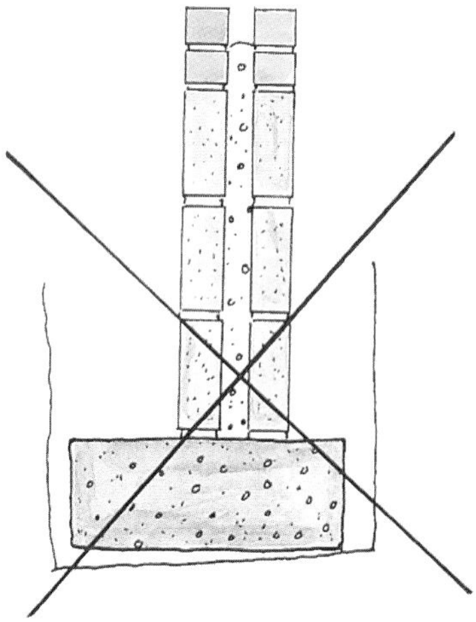

Fig 80 Always build centrally off strip foundations.

Fig 81 Calculate foundation depths using the existing damp proof course as a guide.

52

hand at least one man should mix and pour while another levels off each load.

For larger jobs, pre-mixed concrete from a ready-mix supplier will save a lot of work and speed up the process. The concrete will be delivered to a mix you specify, 1:3:6 is most common for foundations although ground conditions will dictate, and to a working consistency applicable to the circumstances. If the concrete can be poured directly into the footings all will be well, but where the concrete has to be dumped onto a large plastic sheet then wheelbarrowed to the trenches ask the supplier about adding a retarder. This will give you a little 'working' time, usually a couple of hours, before the concrete starts to dry out.

ORDERING CONCRETE

When you order concrete from a ready-mix supplier make sure you tell them what the concrete is for, the area and depth you are filling, and if the load is to be dumped onto a sheet or poured directly into the footings. Do not calculate how much concrete is required too accurately and always allow a little extra for error. If, when the concrete is poured in, you have some left over the lorry driver will usually take it away for you and if you have some left over on the sheet which cannot be used anywhere else on the site, spread it out as thinly as you can and leave it to dry. It can be then used as hardcore or ballast, on the oversite for example. Do not leave it in a huge lump to be broken up later.

WALLS BELOW GROUND

When you have finished the footings and the concrete has had at least a couple of days to set, or 'cure' to be precise, the bricklayer can come in and build the walls up to damp proof level. These walls can be built in blockwork and brickwork, or just brickwork, but make sure that the materials you use, blocks, bricks and cement, are suitable for walls below ground level. Not all materials are suitable for this purpose and will degenerate very quickly due to the extra moisture associated with being below ground. After the foundations have

Fig 82 Level concrete to the top of the guide pegs.

been laid you can use the gauge rod to calculate the number of blocks and bricks required to build the walls up to damp proof course level.

LINTELS

Where drainage pipes pass through the extension foundations, lintels will be required to act as a suitable bridge over

FACING BRICKS

Not all facing bricks are suitable for use below ground level. Check with your supplier that the bricks you are using are suitable for the purpose.

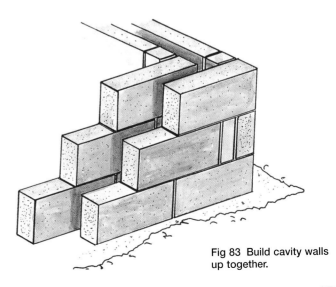

Fig 83 Build cavity walls up together.

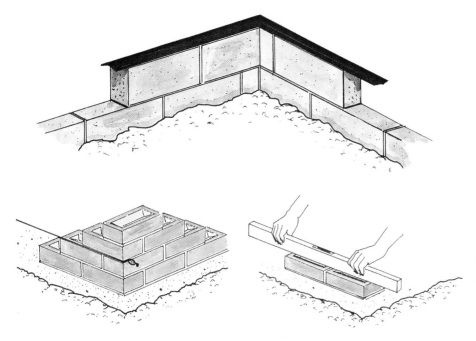

Fig 84 Damp proof course (dpc) should be bedded onto the walls with cement.

Fig 85 *(far left)* When the corners are built the walls can be 'run in' using 'brickies' lines.

Fig 86 *(left)* Make sure walls are built level and upright.

the pipes. Below ground level, pre-stressed concrete lintels are the best option. As with all lintels, an end bearing – that is how much lintel is resting on the walls adjacent to the opening in the wall – of 150mm must be allowed. This will leave openings around the pipe where burrowing animals can dig under the extension. To avoid this cut a rigid sheet to fit around the pipe and to cover the hole in the wall, and secure in place before backfilling the foundation trenches.

UNDERFLOOR VENTILATION

To prevent dry rot and other airborne fungi causing damage to existing and new wooden floors, ventilation must be provided by way of a continuous flow of free air under suspended floors. More applicable for older properties where wooden floor joists and floor boards are common and where the timbers are not, or have

BLOCKS BELOW GROUND

Foundation walling below ground level can be built using Thermalite Trenchblocks. They are lightweight, easy to use, add insulation and do not require a cavity infill, offering a saving on construction costs and time.

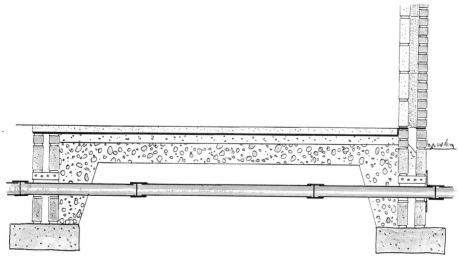

Fig 87 Where drainage pipes pass through walls concrete lintels will be required.

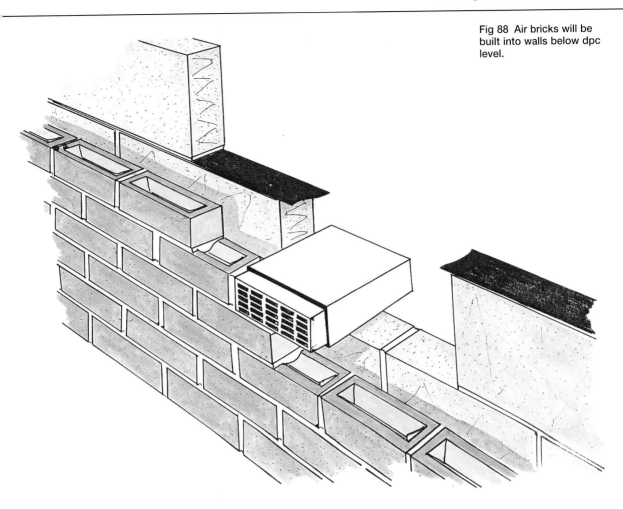

Fig 88 Air bricks will be built into walls below dpc level.

AIR BRICKS

not been, treated against these perils, the existing air bricks providing an opening for this ventilation must be maintained.

Where the air under the floor becomes stale and stagnant, in corners for example, there is always a risk that air conditions will become suitable for such unwanted attacks. If you are installing a timber floor always used timbers treated against such fungal and insect attacks. Precautions must also be taken when building an extension against walls where existing air bricks are positioned to ensure that a free airflow is maintained. Where this occurs new air bricks must be built into opposite external walls and, if the new extension floor is concrete, the air can be ducted into the area at risk by way of pipes laid under the oversite concrete.

Where a new solid concrete floor is to abut an existing suspended wooden floor, it is important that adequate and well-positioned underfloor ventilation is provided. Clay air bricks should be built into the extension walls at up to 3m apart below the damp proof level, and, to continue the airflow through the inner leaf of the cavity wall, a purpose-made clay ducting can be used. In the corners, where stagnant air can become a problem, the air bricks and ductings will be situated within 450mm of the corners. Ventilation pipes will then be laid under the proposed concrete floor from the air bricks to the positions where they can best ventilate the existing wooden floor.

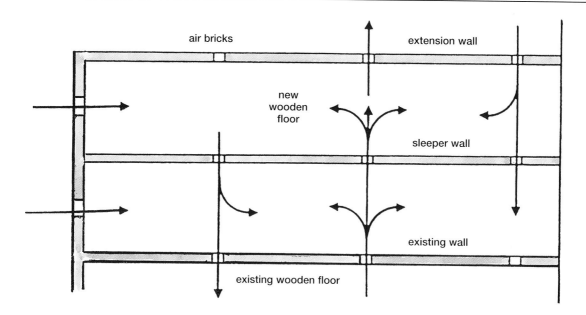

air bricks

extension wall

new
wooden
floor

sleeper wall

existing wall

existing wooden floor

Fig 89 Providing free air ventilation from external walls through to internal walls.

PREPARING THE OVERSITE

When the extension walls have been built up to damp proof course level the area within the walls is called the oversite. You may well have stripped all vegetation off this area at an earlier stage but ensure that all topsoil, vegetative materials and perishables such as wood or roots are removed and the area is cleared ready for the hardcore base to be laid. As a final precaution, and if you

Fig 90 Hire a 'whacker plate' to compress hardcore.

are in any doubt, spray the area with a good weedkiller before starting the next stage.

On this prepared oversite area an even bed of clean hardcore will be laid and then compacted with a heavy compactor or a 'whacker plate' hired from your local plant hire centre. The hardcore should be laid to a minimum depth of 150mm and to a depth not greater than 300mm. Old broken bricks and roof tiles provide the best hardcore and are a manageable size to consolidate without leaving unwanted air pockets. If you are laying a damp proof membrane on top of the hardcore and below the concrete oversite, the hardcore must be 'blinded' with at least 12mm of sharp sand to prevent the membrane being punctured by any sharp hardcore projections.

CONCRETE FLOOR INSULATION

Building Regulations specify that you must reduce any heat likely to be lost through the solid concrete floor. To combat this, a layer of at least 25mm flooring-grade mineral fibre can be installed above the damp proof membrane. This insulation layer can be laid on the membrane but below the

Fig 89 Providing free air ventilation from external walls through to internal walls.

WEEDS
It is imperative that all weeds and vegetative materials are removed from the oversite area before the hardcore is laid. As a safeguard you can spray a strong weedkiller over the surface, making sure to follow the manufacturer's specifications.

concrete oversite, or on the concrete oversite but below the floor finish – concrete screed for example. There are other variations of concrete flooring insulation and if your approved plans specify a variation then this should be followed.

DAMP PROOF COURSE

To prevent dampness rising from ground level and causing damage to both internal and external walls, a damp proof course (dpc) must be installed. The dpc will be fitted to both walls and bedded in with mortar. Where the new dpc abuts any existing walls it will be dressed into the existing dpc to provide a continuous and preventative barrier. The dpc should be at least 150mm above ground level and, where this is not possible, when you are building into a bank or similar higher ground for example, all precautions must be taken to ensure that water penetration to the internal walls is prevented.

DAMP PROOF MEMBRANE

The damp proof membrane is installed to prevent damp rising through the concrete floor. There are several methods in use which prevent this, these include laying membrane sheets on a sand-blinded bed, and liquid bitumen being either painted or laid hot onto the concrete oversite. Firstly, you may consider laying a polythene film membrane sheet, of a quality applicable for this purpose, onto a sand-blinded base. If so you must make sure that all joins are watertight and take the greatest care not to puncture the sheet before and during laying it on the oversite area and again when pouring the concrete.

Secondly, there are several proprietary brands of cold liquid bitumen solutions which can be painted onto the oversite concrete. Manufacturers' instructions must be followed carefully and up to three coats will be required. Thirdly, a hot bitumen, often called tanking, can be poured onto the primed oversite concrete area to a depth of not less than 3mm. Whichever method is chosen it is important to remember that

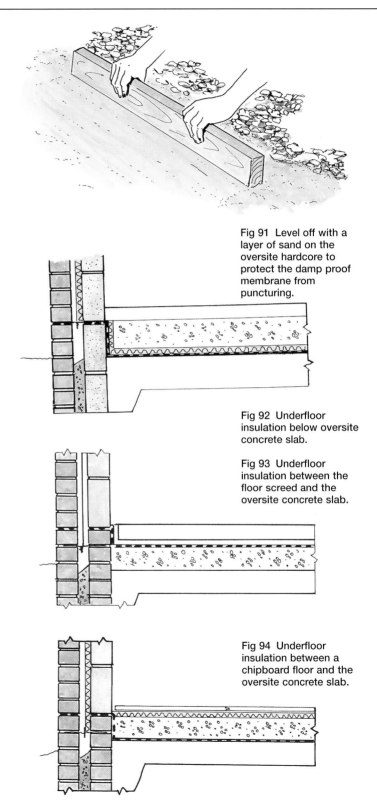

Fig 91 Level off with a layer of sand on the oversite hardcore to protect the damp proof membrane from puncturing.

Fig 92 Underfloor insulation below oversite concrete slab.

Fig 93 Underfloor insulation between the floor screed and the oversite concrete slab.

Fig 94 Underfloor insulation between a chipboard floor and the oversite concrete slab.

Fig 95 Lay the damp proof membrane carefully to avoid puncturing it.

DAMP PROOF MEMBRANE
Great care must be taken when installing a polythene damp proof membrane. Make sure the oversite hardcore is covered with sand and take care not to puncture or tear the membrane when you pour on the oversite concrete.

the floor membrane must be dressed up to, and join, the existing and new damp proof courses to provide an unbroken and continuous protection for the floor area.

LAYING THE OVERSITE CONCRETE

You have excavated the oversite area, laid and compacted the hardcore and, if you are using a sheet damp proof membrane, you have sand-blinded the hardcore and laid the membrane sheeting. Now the next step is to pour the oversite concrete. Oversite concrete is commonly mixed at a 1:2:4 mix, this is slightly stronger than foundation concrete which is commonly mixed at a ratio of 1:3:6. Laid to a minimum depth of 100mm, oversite concrete can be mixed on-site or delivered ready-mixed like the foundation concrete. If you are pouring the concrete onto a membrane the greatest care must be taken not to puncture it, which may result in problems at a later date.

As the concrete is poured it must be levelled and tamped down to remove air pockets and then left to cure for up to three days. Unlike the foundation concrete, where the concrete is unlikely to be exposed to excessive weather conditions, the oversite concrete will be vulnerable to frost and heat damage. The best defence against frost damage is not to lay the concrete at all until the weather permits. If this is not possible then the area must be covered with sacking or similar fibrous sheeting

Fig 96 Level off the oversite concrete leaving a good even finish.

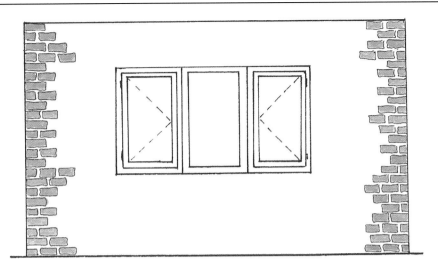

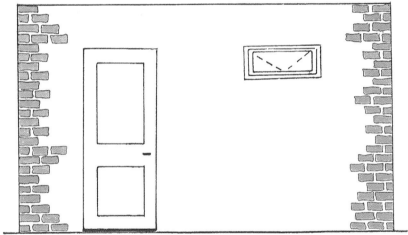

Fig 97 Natural ventilation is very important and will be calculated from opening windows and external doors. An area totalling one twentieth of the floor area may be acceptable. Background ventilation will also be required and can be provided by trickle vents built into window and door frames.

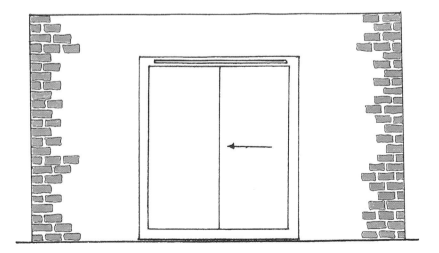

material, leaving a small gap between the concrete and the covering for airflow.

If the weather is hot and sunny, drying the concrete out too quickly then becomes a problem. For the best results the concrete must be kept damp for as long as possible allowing curing to take place over two or three days. Shade from the sun is the best protection with constant, though not excessive, damping down to avoid fast drying out. If you are reinforcing this oversite base, lay a thin layer of oversite concrete, then lay on the reinforcing sheets and finally pour on the top layer of concrete. Speed is important and a continuous flow must be achieved to allow the two layers to bleed together properly.

Important

There are three important points to bear in mind when laying oversite concrete and these are: levels, levels and levels! In order to get the finished floors at the correct level make sure that the level of the oversite base, the level of the hardcore and the level of the oversite concrete are all set correctly.

TRADESMEN

This phase of work will involve the bricklayer, who will build the walls up to the damp proof course level, and labour for mixing and pouring concrete foundations and the oversite.

STEP-BY-STEP SUMMARY

Foundations

Before filling the trenches with concrete, calculate at what level you would like to start building walls. Make a gauge rod from an old length of timber and use a permanent guide such as the existing damp proof course to work from.

Pre-Mix

Measure your trenches and pass them on to the supplier. If access is easy and the concrete can be poured directly into the trenches, then the water content of the mix is not crucial. If the concrete is to be dumped onto a sheet and then wheel-

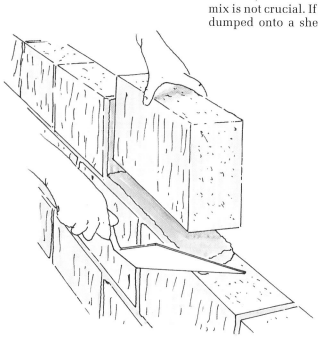

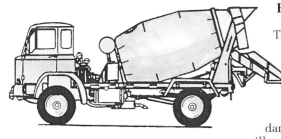

Hardcore

The oversite hardcore should be of mixed sizes without being over-large. Level off and hammer down using a compactor. When the hardcore is down, a layer of sand can be added if you are building in a damp proof membrane. The sand will reduce the risk of tears or holes in the membrane.

barrowed to the trenches, then tell the supplier.

Site Mix

Mixing and laying concrete on-site is very demanding. Make sure the process can be completed in one go, without lengthy breaks, to remove the risk of joins and weak points as it dries.

Oversite

The art of laying a good concrete oversite is to make sure every stage is levelled off. The stripped oversite area, the hardcore, and finally the oversite concrete, must be laid at the correct level to ensure the finished floor level is achieved.

Below Ground

Try to allow a few days before the walls are built off the foundations, giving them time to cure. Make sure all the materials and plant are ready and on-site for the bricklayer. Do not forget the air bricks or concrete lintels as these small items can easily be overlooked.

PHASE 2 – BUILDING (CAVITY WALLS)

With the drainage works completed, the walls built to damp proof level and the oversite concrete laid, you can now back-fill the trenches with rubble and soil up to ground level and generally tidy the site ready for the extension structure to be built.

It is unlikely that you will have a choice of materials from which to build your extension walls, with the Planning Office stipulating that the materials you use match the existing building as closely as possible. Bricks, and old bricks in particular, seldom match exactly and where the original brick makers no longer produce

the same bricks, or are no longer in business, trying to copy the exact bricks may be difficult. When you have to choose a substitute brick it is important to consider what effect weathering will have on it and what the brick will look like in a few years' time. If a near match cannot be found and you decide to use a brick that you believe is complementary to the existing bricks, take a sample to your Planning Department for their approval before you purchase or start any building work.

Bricks are the most popular material used in house building today and can be dated back at least 5,000 years to the

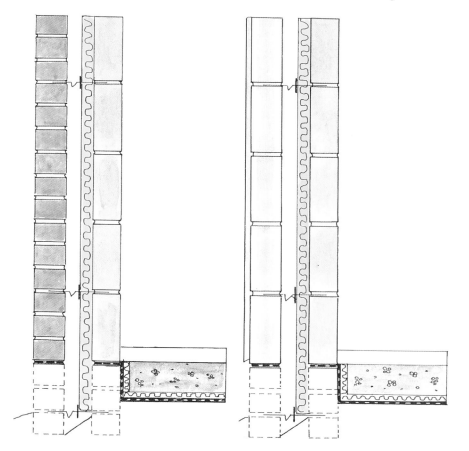

Fig 98 *(Far left)* Cavity insulation in a conventional brick/insulation block wall.

Fig 99 *(Left)* Cavity insulation in a block and render/insulation block wall.

Egyptians, but they are not the only material in use. Natural stonework, for example, dates back to long before the invention of the brick, but this extremely attractive and desirable product has become expensive to use in modern construction and is not easy to build with. To satisfy demand and to retain appearances where existing stone buildings are common, a pre-formed, man-made, block with a stone appearance is used. These blocks are less expensive to build with, readily available and provide a suitable alternative to natural stone.

Other materials in use include concrete blocks with a concrete rendering finish and walls with a roof tile façade. These, and many others, all add to the diversity of choice available to designers. In fact, construction technology used in the modern home has advanced to such an extent during the last fifty years, with damp proofing, cavity wall insulation and other energy-saving innovations playing a major role in wall design, that only minor improvements can be expected in the immediate future.

CAVITY WALL CONSTRUCTION

All the external walls of new buildings, and this includes windows and doors, must meet with stringent Building Regulation requirements before approval will be given. Cavity wall construction, which has been designed to provide protection from damp penetration and reduce heat loss from the building, consists of a decorative outer wall, a cavity, filled or partially filled with an insulator, and an internal wall. Damp penetration is one of the most serious building defects and can cause extensive damage to the structure, both inside and out. When a cavity wall is built properly it should not permit damp penetration and, when it does, it is usually due to the cavity being bridged, often as a result of mortar droppings left on wall ties, or mortar dropping down the cavity during construction.

The other major factor where external walls must comply with current Building Regulations is heat loss. Each wall, including windows and doors, must meet accepted thermal requirements, 'U' values, from which the building's energy use

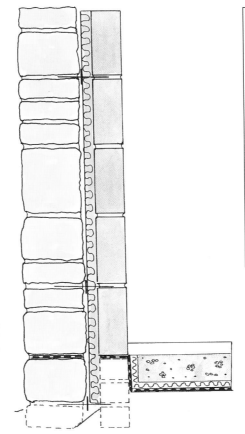

Fig 100 Cavity insulation in a 150mm dressed stone/insulation block wall.

WEATHER
Bad weather can seriously damage walls during construction. Rain can wash mortar out of joints and frost can reduce 'green' mortar to dust in a very short period of time. In extreme weather conditions, brick and block laying should be avoided and all new walls adequately covered when temperatures fall or when it is raining. Sand, bricks and blocks must all be covered during inclement weather.

can be calculated. To assist the retention of heat, the internal wall can be built using insulation blocks with more insulation added to the cavity. There are numerous variations on this theme with the one single aim of keeping heat loss to a minimum and producing a cavity wall which meets all the statutory requirements.

MORTAR

In the dictionary, 'mortar' is defined as a mixture of sand and cement used to join stones and bricks together. In fact it is far more important than that. A good cement mortar is essential to achieve the right finish and lasting strength to your walls. Portland cement and soft sand at a ratio of about one part cement to four parts sand is a widely used recipe for walls, both above and below ground. Non-hydraulic and semi-hydraulic lime can be added for workability and strength and to lighten

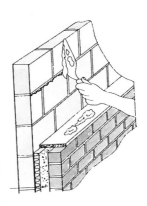

Fig 101 Keep cavities free from mortar droppings.

the mortar, with other additives including a plasticizer, for workability, and a colorant, to add colour.

The mixture should not be too wet, or the bricklayer will struggle to lay it as a bed for the bricks. If it is too weak (too much sand), the joints may crumble away and if it is too strong (too much cement), the joints may crack. Enough water should be added to produce a good dropping consistency off the trowel, allowing it to spread easily and providing enough time for the brick to be laid to a line and level. Where the bricks, the sand or the weather increase the drying out time, making life difficult for the bricklayer, a plasticizer can be added. A good local brickie will know local sand grades, when best to lay the bricks weather-wise, and the strength of mixture most suitable for the bricks you use.

Finally here are three simple points for producing the best brickwork: 1) use an experienced local bricklayer; 2) keep the sand covered when not in use; and 3) store the cement in a dry area, off the ground.

CAVITY WALL INSULATION

With fuel costs rising annually, energy saving is money saving. In England and Wales there are guidelines for walls and windows which must achieve a minimum in energy saving, calculated in 'U' values. The 'U' value of a wall shows the thermal transmittance (heat transfer) through a wall when outside temperatures differ from inside temperatures. To limit the heat loss through the fabric of the building, cavity wall insulation can be built into it in various forms.

Standard forms of insulation include using insulation batts 50mm thick to fill the cavity, and expanded polystyrene sheets 25mm thick attached to the inner wall using special clips. Good site care is necessary to ensure that the installation process is carried out successfully. Where the cavity insulation fills the cavity all mortar droppings should be prevented from falling down the cavity and the insulation batts must be protected against wet weather. Where the expanded polystyrene sheets are used, they must be firmly attached to the internal wall and not left flapping about freely. Installation

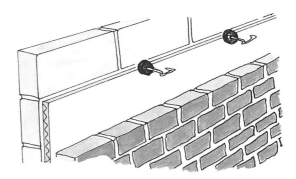

problems may arise where the wall ties are fitted badly or irregularly, causing the slabs to be chopped about unnecessarily.

The cavity insulation will be fitted by the bricklayer as he builds the walls and it must be kept clean and dry at all times. Store the insulation in a dry place until it is used and always cover an unfinished wall overnight to protect it against dampness.

Fig 102 Cavity insulation should be installed according to manufacturer's specifications.

THERMAL BRIDGING

In recent years, the insulation of domestic buildings has come a long way but there is still a weak link, around windows and doors for example, where a 'bridge' between the cold outer wall and the warm inner wall is formed. For instance, on a cold day the warm air inside the house will travel to the colder outside air. This is called thermal conductivity and is caused by a drop in temperature. It increases the risk of condensation and mould growth. For the purpose of calculating the insulation

PLASTICIZER
Using a plasticizer in mortar will add air and help protect the mortar against some low-level frost damage, but it will also delay setting times.

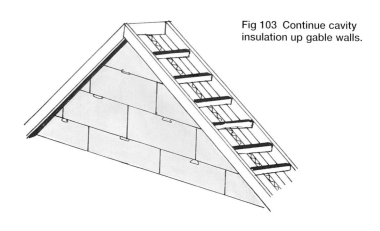

Fig 103 Continue cavity insulation up gable walls.

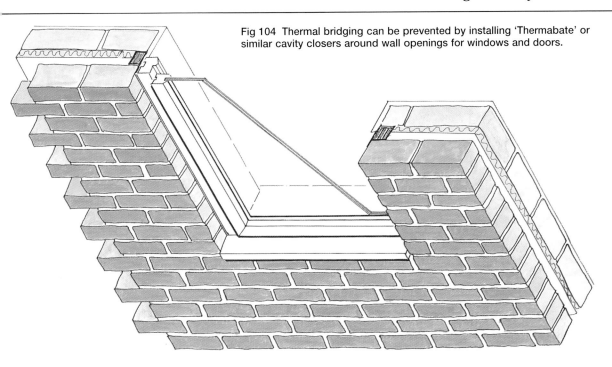

Fig 104 Thermal bridging can be prevented by installing 'Thermabate' or similar cavity closers around wall openings for windows and doors.

potential or resistance of fittings in close proximity to a bridge, a chart showing the 'R' values, or resistance values, is used. To achieve the required 'R' values, and to meet Building Regulation requirements, lintels, because they are often situated in the direct vicinity of a 'bridge', have been filled with insulation to improve their respective 'R' values. This has been done so that when you purchase them, you can accurately compare the 'R' values of one brand with those of other brands.

Where the cavity is closed in other areas, reveals around windows and doors for example, thermal bridging products incorporating a dpc can be installed to reduce the risk of condensation and mould.

VERTICAL DPC

Damp penetration from the ground upwards can be prevented with the fitting of a damp proof course. The same method of protection can also be applied to window and door openings in walls. Where the external and internal walls are joined together, at the 'reveals' around windows and doors, water penetration is possible, so a vertical damp proof barrier

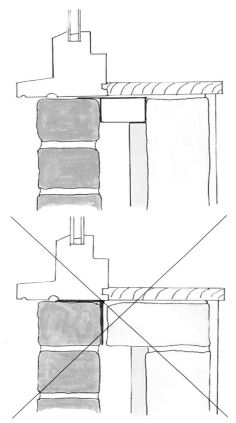

Fig 105 *(Left)* Cavities closed around windows and doors using special closures.

CAVITY CLOSERS
A cheaper method of closing cavities around windows and doors is the use of 'Thermalite' coursing bricks. These bricks are specially designed for this purpose and can also be used for split courses and in-filling between joists.

Fig 106 How not to close cavities around windows and doors.

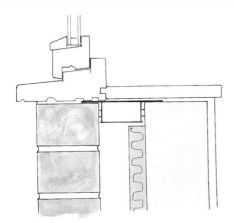

Fig 107 A thermal bridge cavity closer.

Fig 108 Installing insulated vertical dpc.

must be installed during construction as a preventative measure. An insulated cavity closer can be used which will tackle both the problems of damp penetration and heat loss, through thermal bridging, at the same time.

WALL TIES

To build a stable, strong cavity wall, wall ties must be fitted at regular intervals. There are several types on the market, all designed to add strength to cavity walls in particular, and to restrict water penetration where the wall tie forms a bridge connecting the outer wall to the inner wall. Poor installation, however, where a wall tie is not level and slopes toward the inner wall, can provide the bridge necessary for rainwater after it

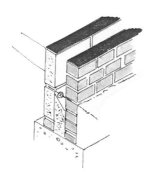

Fig 109 Set the first row of wall ties at ground level.

Fig 110 *(Above)* Wall ties must be installed correctly.

Fig 111 Install wall ties at regular intervals.

has soaked the outer wall, transferring the dampness to the inner wall. Your choice of wall tie may well be dictated by the type of cavity insulation you use, but with all of them the wall ties must be kept clean and free of mortar droppings during construction.

INDENTS AND PROFILES

Where your new extension wall abuts the existing house wall, a proper join must be made to ensure stability. There are two principle ways of carrying out this process: The first is by cutting indents into the existing wall, forming a mortise and tenon-type joint, and then building the new wall into the indents.

Fig 112 A vertical dpc is required where any part of an external wall becomes an internal wall.

Fig 113 Maintain the dpc protection by dressing the new dpc into the old dpc. A vertical dpc may also be required.

The second is by fixing profiles against the existing wall and building the attachments in as the wall is built. This connection will also provide a 'bridge' where dampness can penetrate through to the inside walls so a vertical dpc must be fitted. Using a 'disc cutter' or 'angle grinder' hired from your local hire shop, cut a line through the wall you are building into, approximately central to the cavity of the new adjoining wall, for the full height of the building. A dpc can then be inserted to provide a continuous barrier against damp.

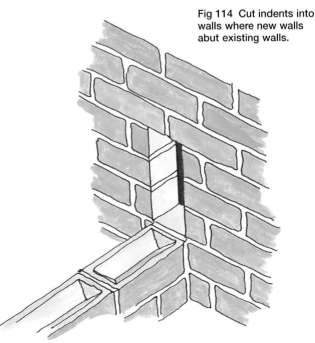

Fig 114 Cut indents into walls where new walls abut existing walls.

LINTELS AND ROLLED STEEL JOISTS (RSJs)

A lintel is a horizontal beam, originally made from stone and forming the shape of a brick arch, but now more commonly seen in galvanized steel or pre-cast concrete. The lintel is built into a wall as it progresses and provides a support for masonry above an opening such as a window or a door. To function properly the lintel must be installed correctly and should not be damaged in any way prior to installation, as this damage may result in the lintel failing in its task. The lintel will be built into the new wall as it progresses and it will be bedded on mortar with both the inner and outer walls being built up together.

For lintels, Building Regulations are particularly clear about *three* specific points:

1. The lintel must have an end bearing, rest on each supporting wall, with at least 150mm at each end.
2. The thermal bridging requirements ('R' values) must be achieved.
3. The water penetration, a result of water travelling down the cavity, must be directed to the external wall.

To satisfy points 2 and 3 lintel manufacturers have amended the design of their product so that the lintels are not only insulated to meet these requirements but also designed, like a cavity tray, to direct downward water to the outside wall. To assist this final point, additional precautions must be taken to allow this water to exit through weep holes built into the external wall at either end of the lintel, including the proper installation of stop

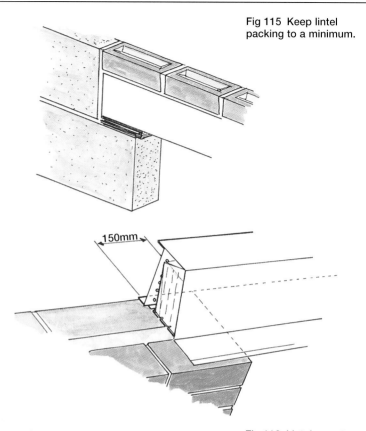

Fig 115 Keep lintel packing to a minimum.

Fig 116 Lintels must have a minimum of 150mm bearing on supporting walls.

ends to direct moisture toward the weep holes.

An RSJ, a rolled steel joist or Universal Beam, may be specified where extra large openings in walls are considered

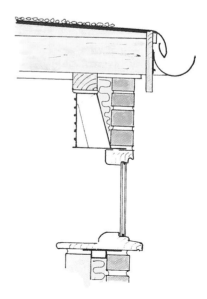

Fig 117 Section showing cavity wall and lintel.

PIERS AND POSTS

Sometimes support for a lintel or RSJ is required in a position where a wall would be unsuitable. In these instances a pier or post may be the best solution as they provide the required support without taking up the room required for a wall. Depending upon the load to be carried by the pier (it may be walling, flooring, or roofing), it is likely that it will require a proper concrete foundation, dug to a required depth and inspected and approved by the Building Inspector.

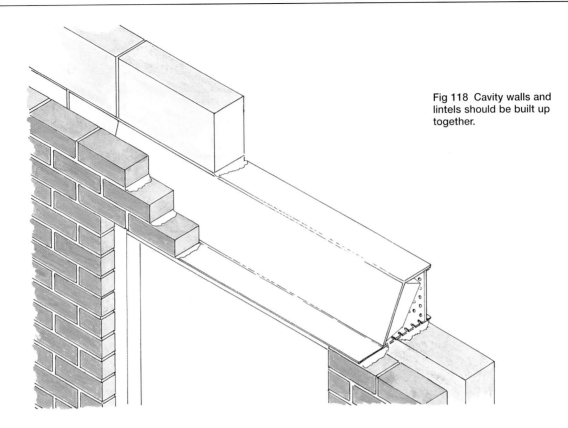

Fig 118 Cavity walls and lintels should be built up together.

which fall outside the scope of standard lintels. These strong beams can be purchased in a variety of lengths and cross sections appropriate for almost any circumstances. Where a beam is required it is likely that Building Control will request structural calculations to prove the ability of the beam to carry the load to be imposed upon it. Unlike standard steel lintels these beams can be extremely heavy, putting an extra emphasis on installation. The beam must be bedded onto special load-bearing padstones or engineering bricks, and in some cases steel plates, in order to spread an imposed load, with the potential to crush standard bricks and blocks, over a wider area.

Fig 119 (A) An RSJ supporting timber joists with noggins to provide stability. (B) An RSJ boxed in timber frame with double thickness of plasterboard to provide fire protection. (C) Joists supported by an RSJ. (D) Joists supported by timber plate attached to an RSJ.

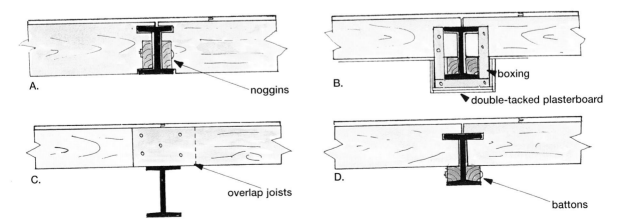

A.

noggins

B.

boxing

double-tacked plasterboard

C.

overlap joists

D.

battons

DOOR AND WINDOW FRAMES

One of the more important decisions you make during the planning process is to select the style, size and positioning of door and window frames. The Planning Department will also take particular care that any windows and doors meet with their specifications, which include privacy for neighbouring properties and

Building Regulations stipulate a minimum area of window opening for any habitable room to be at least one twentieth of the room's floor area, plus background ventilation, with a proportion of the primary ventilation at least 1.75m above floor level. Background ventilation, generally fitted to windows in the shape of a trickle vent, will provide a source of constant and additional ventilation acting as an aid to reduce the risk of condensation.

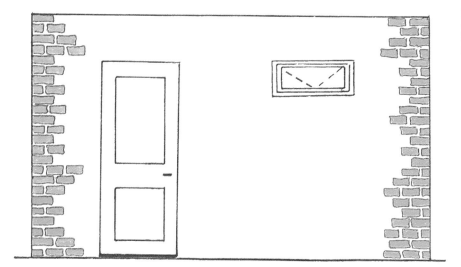

Fig 120 External doors and windows will be included in the ventilation calculations.

Fig 121 *(Below)* Softwood window frames should be primed before installation.

proximity to boundaries. Building Control, on the other hand, consider windows to be a recognized source of potential problems relating to light, heat loss and ventilation. A door frame, without the door, will not encounter the same degree of inspection at the plans stage as that applied to windows, but there are rules for doors that must be followed, depending upon the location, and these include fire precautions.

The range of windows currently available from suppliers is vast, with a multitude of styles to choose from including top hung, side hung, sash and tilt windows. The glazed area of each window is important because it will let in the sunlight you require but it will also let out heat. Double and triple glazing is available as an energy saving option if required.

New bathrooms and toilets attract the closest inspection where you will find ventilation is high on the list of priorities.

UNGLAZED WINDOW FRAMES

It is standard building practice for brick-layers to build unglazed window frames into walls as they progress. Before installation, wooden frames, including freshly sawn areas, should be stained or painted with primer and undercoated, then covered with sheeting as a protection against mortar damage. Frame ties, galvanized or stainless steel are the most common, will be screwed into the frames at regular intervals and then bedded into the mortar joints as the walls are built.

WINDOW TEMPLATES

Building new 'made-to-measure' windows, with the glazing installed, into the walls as they progress can be a very risky

Fig 122 *(Above)* Build window and door templates into walls during construction.

TEMPLATES
Building expensive windows and doors into walls during construction may result in damage to the frames. To prevent damage, templates can be made and built into walls, then removed when the walls are completed and the new frames fitted.

Fig 123 Make templates for specialist door and window frames.

practice indeed, with a serious possibility of damage. To avoid such catastrophes and enable the work to progress without a hitch you will find the best and most popular solution to this problem is to make, or get your carpenter to make, templates for the bricklayer to build the new walls around. These templates can then be removed and discarded and the new windows installed when all the construction work is completed. Each template can be made according to the sizes specified by your window manufacturer, but a strong word of caution: the templates must be accurate; trying to fit a window into an opening too small for it can pose massive unwanted difficulties.

DOOR FRAMES

Door frames, with and without a sill, are commonly made from 100mm × 75mm softwood with a rebate for the door. As with wooden window frames, door frames can be built into the walls as they progress. Frame ties at regular intervals, at least three on each side, will anchor the frame into the mortar joints. Entrance doors, front and rear, invariably open into the building, whilst doors opening onto a balcony or garden area will open out. To prevent rainwater penetration, a galvanized water bar will be fitted to the hardwood sill set at a point approximately central to the bottom of the door. When the door is fitted the bottom can be rebated over the water bar as required. All timber frames should be treated or primed and undercoated before installation, with the hardwood sill doubly protected and covered against mortar or traffic damage.

SCAFFOLDING

The technical term used in the trade when referring to the height of walls during construction is the 'lift'. A 'lift' is the height at which scaffolding is required for tradesmen such as bricklayers to work from. For example, when a wall reaches the first 'lift' height the bricklayer will require scaffolding to stand on so that the building can be built up to the next 'lift'. Similarly, when the walls are built up to the second 'lift' then a further level is required to put the roof on. The scaffolding referred to in this example is called 'pole' scaffolding where poles, or standards, are raised vertically and lateral poles, called ledgers, are bolted on horizontally.

There are two types of pole scaffolding commonly in use. The first is a 'putlog' scaffold where special short poles with one end flattened off, called putlogs, are built into the walls as a support for the planks the builders will work off. The whole scaffold is then braced at regular intervals through openings in the wall, windows for example, for stability. The second type of pole scaffold is an 'independent' scaffold, an almost free-standing scaffold, where two rows of standards are raised vertically and then bolted together with ledgers. This second type may appear to be free-standing but all scaffold must be tied into the building at regular intervals. Pole scaffolding of this type must be erected by highly skilled professionals as there are very stringent safety measures to abide by.

SCAFFOLD HIRE

The height of your finished extension will determine the type of scaffold you will require for the work to be carried out

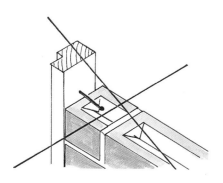

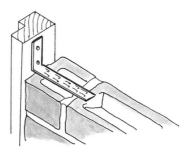

Fig 124 Build frames securely into walls using frame ties.

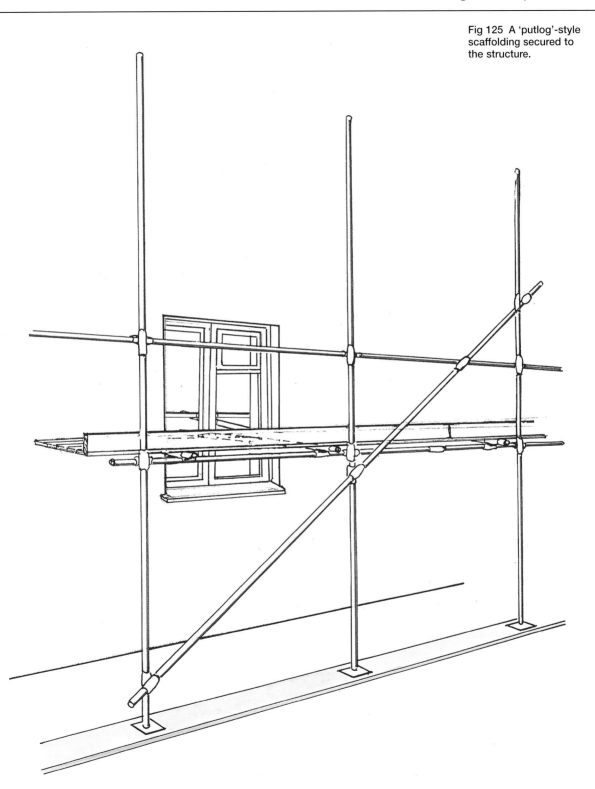

Fig 125 A 'putlog'-style scaffolding secured to the structure.

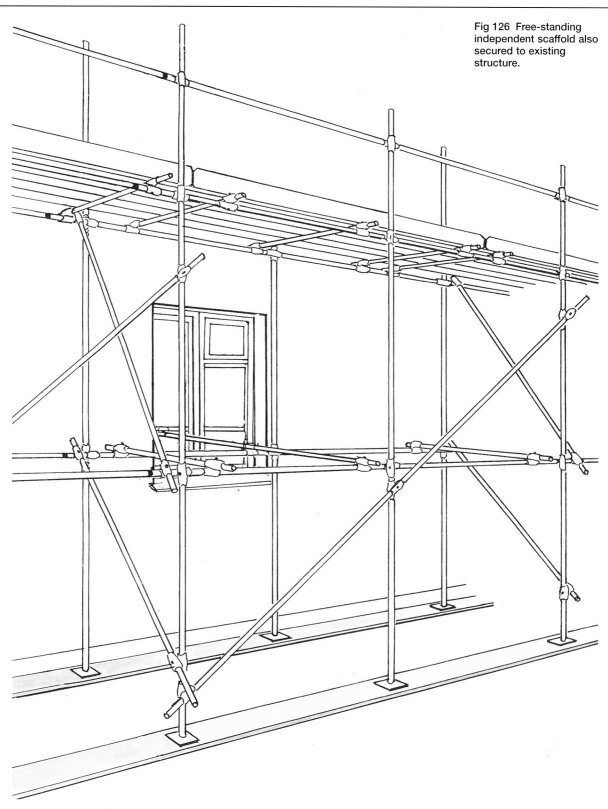

Fig 126 Free-standing independent scaffold also secured to existing structure.

safely and successfully. If the height is under 2.5m then trestles and planks are an option. These can be hired from your local hire shop and erected by the bricklayer during construction. If the extension height exceeds 2.5m, or if the tradesmen request it, then pole scaffolding is likely to be the safest option. Professional supply and fit scaffolders are a must and they can be found in your local Yellow Pages. They will quote you for scaffolding the whole project, including additional lifts, then deliver and erect the lifts as required. Needless to say, close organizing with the scaffolders and the tradesmen will go a long way to ensuring that the project can continue without delays.

Fig 127 Take precautions to prevent items falling from the scaffolding.

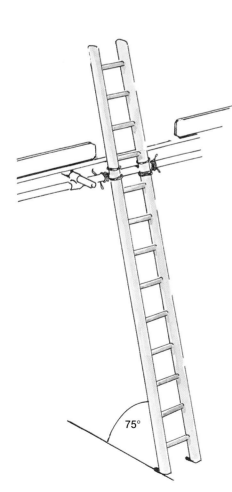

Fig 128 Ladders must be secured to scaffolding and set at an angle of about 75 degrees.

WALL PLATE

When the walls reach roof level, be it single-storey or two-storey, a 100mm × 50mm softwood wall plate, to rest the roof timbers on and secure the roof to, will be fitted. In the majority of cases the wall plate will be

Fig 129 Secure wall plates to internal walls using steel straps.

bedded on the inner wall and then secured with mild steel galvanized restraint straps set at a maximum of 2m centres and starting no more than 450mm from the corners. To ensure that the roof can be built geometrically, the wall plates, bedded onto opposite walls, must be level and squared with each other.

BRICKLAYING

One of the oldest building trades in the business, the bricklayer is responsible for the most visible part of your new building. For this reason alone, hiring an experienced local bricklayer is essential. The 'brickie' will quote you for the job as a whole, or to carry out the work on a day rate basis according to the degree of difficulty involved. A common home extension will generally be quoted on a fixed price and will include setting out the walls at foundation level, building the walls, including cavity insulation, up to roof level, building in door and window frames where required and bedding on the wall plate. The quote will not include supplying materials or plant hire. Nor is it likely to include, unless otherwise agreed, making templates for window units, erecting pole scaffolding or erecting gable end rafters where gable end brickwork is involved.

A good working relationship with all the tradesmen is required and there must be a clear understanding about what is, and is not, included in the price. All the materials must be discussed and they must be available on-site when the bricklayer arrives to start.

TRADESMEN

This phase of work is likely to involve the following:

- The bricklayer, the carpenter and the scaffolder.
  - The bricklayer: will build the walls up to wall plate level.
  - The carpenter: will build templates if required.
  - The scaffolder: will erect scaffolding accordingly.
- Plant hire: scaffolding, cement mixer, angle grinder.
- Labour: will cut indents into walls or fix profiles.

STEP-BY-STEP SUMMARY

Walls

Cavity walls can be built very quickly indeed. Make sure all the materials are on-site and that the bricklayer understands the plans and is familiar with your requirements. It may be too late to change things at a later stage and you will not want to come home from work and find a wall built in the wrong place.

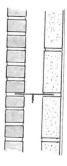

Templates

Window and door frames may be built into the work as it progresses or they may

---

QUOTATIONS

Bricklayers are generally paid by the job, called 'price work', or by the quantity of bricks and blocks laid, called 'measured work'. Small home extensions are normally classified as price work, requiring one payment, in full, when the bricklaying is finished. For larger projects interim payments may be requested.

Included?
Apart from building the walls as specified on the plans, it is reasonable to expect that the bricklayer's price will include, unless otherwise agreed:

1. Setting out the walls on the foundations.
2. Building in door and window frames or templates.
3. Bedding on the wallplate.
4. Installing the cavity insulation.

Not Included?
It is also reasonable to expect that the bricklayer's price will not include, unless otherwise specified:

1. Making templates for windows and doors.
2. Erecting rafters for gable end work.
3. Erecting pole scaffolding.
4. Supplying materials.

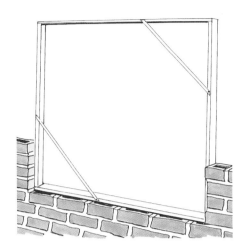

Indents

Joining new walls to existing walls will require additional work. Check with the bricklayer what is, and is not, included in the price and make absolutely sure that work not included, such as indents and vertical dpcs, has been done before they start work.

Scaffolding

When scaffolding is required, and is to be erected by others, make sure both the bricklayer and the scaffolder are aware of the situation. Avoid unnecessary hold-ups if you can.

be added later. If templates are to be used, make sure they are ready and on-site and that the bricklayer fully understands which template goes where.

Are You Prepared?

The bricklayer will want to finish as quickly as possible in order to move on to the next project. He will not want to return to build a small gable end wall, or to bed on wall plates. It is imperative that all the materials required are available and on-site. Leave a contact telephone number in case of emergencies.

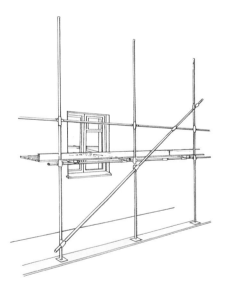

Weather

It is possible to build walls in inclement weather. Cold days and nights can become a problem along with heavy rain, but in the majority of cases conditions should be alright. In the event of overnight frosts and sudden downpours, make sure sheets are available to cover and protect new walls and cavity insulation.

Double-check

Always remember that you are totally familiar with your project, but to the bricklayer it is just another extension of many. He cannot be expected to know of any changes or additions unless he is told. Make sure he fully understands your requirements. When a bricklayer makes a mistake the rectification work is usually major.

PHASE 3 – ROOFING 1 (FLAT ROOFS)

To many observers the completion of a roof will spell the end of the building work but, in fact, this is hardly the case. True, the completed roof construction generally signals the end of all the major works which include excavations, drainage works and cavity walls, but it also signals the start of the finishing trades which may well take at least the same, if not more time, to bring the project to completion.

For tradesmen there is likely to be a change of personnel as the carpenter replaces the bricklayer and then the roofer replaces the carpenter. At this stage it is important that any hand over from one trade to another goes without a hitch. And by hitch I mean that the building is built squarely and level, the timbers are secure and the scaffolding is stable. When a roof is fitted the carpenter will soon discover if the building is square, or not, and how level the wall plates are. For simple roofs, such as a flat roof, the variations may not be too difficult to live with but for more complicated roofs, hipped roofs for example, a building out of square could prove to be disastrous, and expensive to rectify.

FLAT ROOF CONSTRUCTION

Historically, the flat roof has been widely used in the Middle East where the climate is dry and arid for many centuries, but it wasn't until the nineteenth century and the introduction of waterproofing techniques that the flat roof became more widely used in Europe.

Possibly the simplest of all roofs to install, a flat roof is extremely popular on single-storey extensions, though not usually allowed for buildings above one storey. A flat roof consists of timber joists resting on, and secured to, wall plates. Layers of bitumen felt are built up and laid hot on a sheet decking to provide a waterproof covering, with insulation laid on top of, or between, the joists to ensure that the structure achieves the energy saving requirements necessary to comply with Building Regulations.

There are two types of flat roof: a cold roof and a warm roof. The cold roof is constructed with the insulation fitted in the roof void between the roof decking and the ceiling boards. To avoid condensation problems occurring in cold roofs the void, the area between the joists, must be ventilated with a clear air flow above the insulation and below the roof deck, assisted by ventilation gaps along the sides of the roof. A warm roof is constructed with the insulation built in between the bitumen felt waterproof covering and a vapour barrier fixed to the decking.

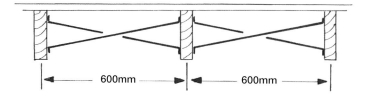

FLAT ROOF JOISTS

Timber joists, spanning from one wall to an opposite wall, will be laid level and and at agreed centres, ready for the flat roof decking to be attached. There are three factors to consider when ordering your roof joists: the size, grade and length. The joist sizes, 150mm × 50mm, 175mm × 50mm and 200mm × 50mm, are common and will be specified on your plans. The grade, or timber strength, may also be specified and will be one of three common grades of timber used for this type of construction: General Stress (GS) grade, Special Stress (SS) grade and Machine General Stress (MGS) grade. The GS and SS grades are assessed visually, noting the grain and the position of knots and so on, and the MGS grade is assessed mechanically.

Fig 130 Set joists out at the centres (600mm for example) specified on the approved plans.

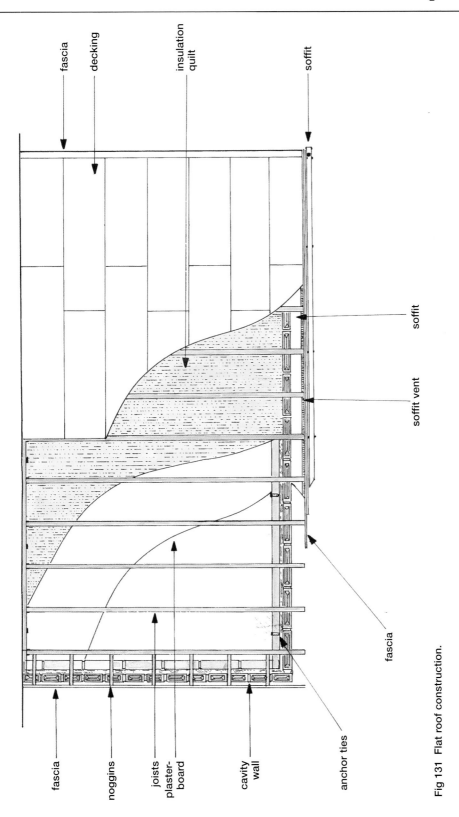

fascia

decking

insulation quilt

soffit

soffit

soffit vent

fascia

fascia

noggins

joists

plaster-board

cavity wall

anchor ties

Fig 131 Flat roof construction.

The length of the joists may be calculated from the plans but will be better calculated from the building itself, allowing a little extra length for tolerances and installation.

Each joist will be fitted to the wall plate at a specified 'centre' as shown on the plans. The 'centre' measurement is taken from the centre of one joist to the centre of the adjacent joists, and so on. You may reduce the centres, to accommodate the decking sheets for example, but you should not increase the centres without prior discussion with the Building Inspector. The joist sizes will be calculated by the length of span, from one wall to another, and the load imposed upon them, in this case the felt roof. When the joists are in position and firmly nailed to the wall plate, mild steel restraint straps will be fitted. The restraint straps should be fitted at approximately 2m centres and within 450mm of corners. They will be securely fixed to a joist and will extend down the wall at least 1m. These straps will then be firmly secured to the wall with at least four fixings, galvanized nails or screws are recommended, with the bottom fixing no more than 150mm from the bottom.

FIRRING SECTION

A flat roof is a slightly misleading term because, in fact, the roof is not really flat. If it were then rainwater damage would reduce its life substantially. To provide the necessary slope on a flat roof, and to direct the rainwater away and into the gutters, a firring piece will be fitted to each joist before the decking is laid. A firring piece is a strip of timber which can be ordered already machine cut by your supplier, generally the same width as the joist, and to provide a fall, or slope, of at least one in forty.

CONDENSATION

Condensation is formed where hot air meets cooler air, in the roof void for example. Large quantities of condensation are produced in the average modern home from cooking, taking a bath or having a shower. Even breathing and

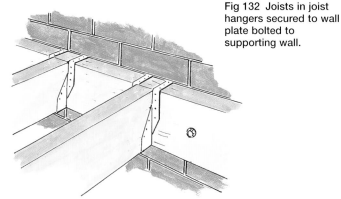

Fig 132 Joists in joist hangers secured to wall plate bolted to supporting wall.

perspiring produces condensation. This warm, moist air will be drawn by convection to parts of the building where the air is cooler. Modern homes are now so well insulated in walls and windows that the roof void is left as a very vulnerable area. Once the warm air has entered the roof void, condensation can occur. Cross ventilation of this area will remove the warm air before damaging condensation can occur.

CROSS VENTILATION

To reduce the risk of damage to timbers and roof fixings as a result of condensation, and to prevent mould growth within the roof void, an effective roof ventilation system is required. Free air must be allowed to flow from one side of the roof to the other without interruption, above the layer of insulation and below the flat roof decking. To achieve this, a gap at least 50mm wide will be left between the roof insulation and the roof decking and this will provide a good space for ventilation over the bulk of the roof, supported by continuous ventilation along the eaves. This continuous ventilation will be provided by openings along opposite sides of the roof. The openings can be a continuous gap along the eaves where eaves ventilators can be fitted or, alternatively, circular eaves ventilators can be fixed into the soffit board (see Fig 163).

Whichever system you choose, the combined areas of the opening must be equivalent to a continuous gap of 25mm running the full length of the eaves.

BUILDING REGULATIONS

In the majority of situations, ventilation for your flat roof will be provided by two processes:

The first will provide an inlet for the outside air to reach the internal roof void, where condensation can occur.

Eaves ventilation is a common method with precautions taken to prevent birds and insects entering and nesting in the roof void.

The second will be a layout within the roof void to allow this air to flow freely between the insulation and the roof decking. Flat roofs, defined as a roof with a pitch of less than fifteen degrees, should have a continuous gap of 50mm between the insulation and the roof deck.

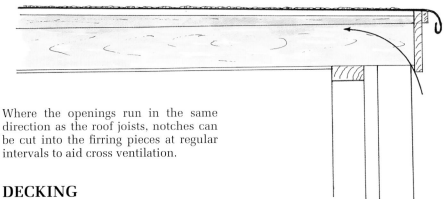

Fig 133 Provide continuous flat roof ventilation at eaves level.

> **WEATHER WARNING**
> The flat roof decking should be fitted just prior to the felt roofers laying the roof. It is important that the decking is not exposed to heavy rain so protective sheets should be available should such conditions prevail.

Where the openings run in the same direction as the roof joists, notches can be cut into the firring pieces at regular intervals to aid cross ventilation.

DECKING

Sheets of exterior grade plywood or tongued and grooved chipboard will provide a stable 'deck' onto which the hot bitumen felt or, if you are building a warm roof, the insulation, can be laid. With the roof joists centred to reduce wastage and to provide support for the sheets where joins occur, the decking can be nailed to the roof joists using galvanized nails. The thickness of the plywood or chipboard sheets you use will be determined by the space between the joists.

TILT FILLET

A tilt fillet is a triangular-shaped piece of timber used around the perimeters of flat roofs to provide a guide to deflect rainwater into the gutters. The tilt fillet will be nailed to the decking, along the edges where required. The bitumen felt, finished in a layer of mineral felt roofing, will be dressed over it for protection.

FELT ROOFING

Asphalt, bitumen-based waterproofing methods have been in use for many centuries and records can be found as far back as the rebuilding of Babylon and more recently, 200 years ago, when it was applied to ships' timbers to make them waterproof and more seaworthy. A natural product, asphalt can be produced from limestone rocks and shale, and in some areas is found in a liquid form. This natural asphalt is not found in the UK but is common in France, Switzerland and Germany where asphalt-impregnated limestone is mined. The largest natural, liquid form of asphalt deposit can be found in Trinidad, known as Trinidad Asphalt Lake, and it has a consistency almost dense enough to walk on.

Recent years have seen great improvements in the durability of asphalt resulting in widespread usage throughout the

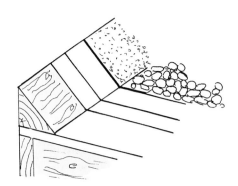

Fig 134 Tilt fillet edging to a flat roof.

> **FELT ROOFERS**
> Failure of a felt roof will have a hugely damaging effect on many other areas of your extension. Roof insulation, ceilings and electrical fittings will all be affected if the felt roof fails. For this reason it is absolutely vital that you use a known, reputable company experienced in this kind of work.
>
> *Not Included*
> Unless otherwise agreed the felt roofers will not:
>
> 1. Insert a cavity tray;
> 2. Fit a lead flashing.
>
> *Quotations*
> Quotes for felt roofing will be based upon the area to be covered and calculated by the square metre. Add to this the edging work involved for the upstands around the roof edges, and welted drip into the gutters, which is calculated by the metre run.

building industry. Mastic asphalt and bitumen-based roofing felt have been combined to produce a highly efficient and commonly used waterproof membrane for flat roofs in particular. The built-up bitumen roof, the most popular form of multi-layer roofing, has gone through rigorous testing over recent years, resulting in the UK in a British Standard CP 144 Part 4, covering the application of mastic asphalt roofing, and BS 747, covering bitumen-based roofing felts, being enforced.

There are two forms of built-up roof, the first used on cold roofs and the second on warm roofs. When applying a built-up bitumen roof covering to a cold roof, the first layer of felt is secured to the deck using hot bitumen or large-headed galvanized nails followed by at least two layers of felt bonded in hot bitumen. Importantly, the additional layers must have staggered joints with an overlap of at least 50mm where joints occur.

To finish off, a layer of reflective mineral felt will be fitted to exposed areas and a layer of reflective stone chippings will be bedded in bitumen to protect the roof from the effects of ultraviolet radiation.

For a warm roof the waterproof covering is laid on top of the insulation, but first a vapour barrier must be bedded on top of the decking for the insulation to be laid onto. A perforated layer of felt is then laid on top of the insulation with hot bitumen poured over the top, sealing it to the insulation through the perforations. A top layer, as with the cold roof, is bedded on and then the process is finished in the same way as with a cold roof. It is important to avoid any traffic movement at all on the completed roof which can cause punctures and result eventually in a breakdown of the waterproof membrane.

CAVITY TRAYS

Water penetrating walls as a result of driving rain, for example, can travel through to the inner wall resulting in unwanted damp patches. To prevent this happening a cavity will be installed in high-risk areas. Lintels over windows and doors now have a built-in cavity tray, but in other areas where an external wall becomes an internal wall, in an extension for example,

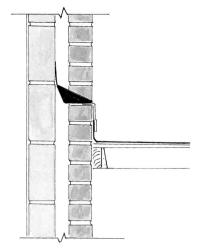

Fig 135 Install a cavity tray above flat-roof positions.

HEALTH WARNING
Never walk under a flat roof when it is being felted. The scolding tar can, and often does, fall between gaps and cracks and can cause nasty burns to the unwary.

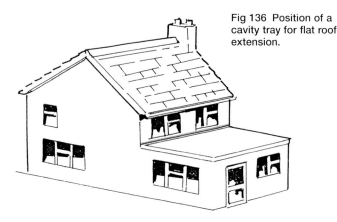

Fig 136 Position of a cavity tray for flat roof extension.

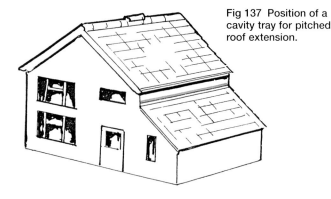

Fig 137 Position of a cavity tray for pitched roof extension.

damp must not be allowed to travel down the cavity and enter the building.

A cavity tray will prevent this and divert any water that has penetrated the outer wall back out through weep holes in the outer wall. Where a cavity tray is required above a flat roof, install it before the built-up felt is laid. Using the roof decking for support you will need to remove a course of bricks, three or four at a time, to insert the cavity tray according to the manufacturer's instructions. Weep holes will be left in the mortar joints at regular intervals to let any water in the cavity escape.

LEAD FLASHING

Where a felt roof abuts a wall, against the existing house perhaps, the mineral felt must be dressed up the wall and into a chased out mortar joint and then re-bedded in mortar. To complete the protection of this joint, a lead flashing can be dressed down over it to finish not closer than 75mm to the roof surface. When a cavity tray is installed it should discharge any moisture from the cavity out over the flashing. This will complete the protection of a vulnerable area where water penetration can be a problem if the process is not carried out in sequence and to a good building standard. The lead flashing specifications will be indicated on your building plans.

FIRST FLOOR JOISTS

Floor joists can be installed at ground floor and first floor levels to provide support for a suspended floor. The floor joists will span between the closest supporting structures and will be built into walls or hung in joist hangers at centres specified on your building plans. The first floor joists will be built into the cavity walls during construction at a

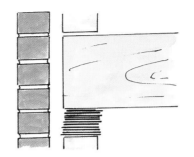

Fig 138 Do not use excessive packing when installing floor joists.

Fig 139 Build first floor joists into walls to provide extra stability.

position where you would normally fix a wall plate for a single-storey extension. The joists will be set out in the same way as you would for the flat-roof joists but, instead of the joists being secured to the internal walls with restraint straps, they will be built into the inner walls as the building progresses. The floor joist sizes, grades and centres will all be specified on your building plans. When you install the joists make sure the end built into the wall is treated with a protective timber treatment. Do not fix the floor decking to the joists until the roof is watertight.

HERRINGBONE STRUTS

To stabilize the timber joists, flat roof and flooring joists, a row of herringbone struts will be installed midway along the length of the joist. If, however, the

Fig 141 Metal herringbone struts provide rigidity to floor and roof joists.

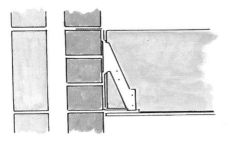

Fig 140 Joists must fit snugly into joist hangers.

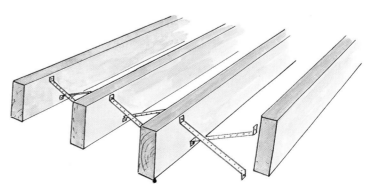

span exceeds 4m, two rows of struts may be required. The struts can be cut from timber or they can be bought from the supplier and securely nailed in place.

Where the joist run ends adjacent to a wall, for example, timber wedges can be installed to provide an overall and continuous rigidity to the whole floor.

TRADESMEN

This phase of work is likely to involve the bricklayer, the carpenter, the scaffolder and the felt roofer.

Single-Storey Extension

- The bricklayer: builds the walls up to wall plate height and beds on the wall plate ready for the flat-roof joists.
- The carpenter: builds the flat roof ready for the roofing felt to be laid.
- The scaffolder: unless you use a trestle-type scaffolding, the scaffolder will erect a first lift for the bricklayer, carpenter and felt roofer to work off. As soon as these trades have finished, unless you have a further need for it, arrange for the scaffold to be removed.
- The felt roofer: will waterproof the flat roof area.

Fig 142 For single-storey extensions and internal works, trestles provide a good scaffold to work off.

- Labour: will install the cavity tray and lead flashing.
- Plant hire: scaffolding and a disc cutter to install the cavity tray.

Two-Storey Extension:

- The bricklayer: will build the walls up to first floor level for the floor joists to be installed, then build up to wall plate level ready for the roof timbers.
- The carpenter: will oversee, if required, the floor joist installation.
- The scaffolder: will build a first lift for walls up to, and past, first-floor level and will raise the level again as a second lift brings the walls up to roof level.
- Labour: will assist with floor joist installation and possibly chasing out holes or bolting plates to walls for the floor joists.
- Plant hire: scaffolding.

STEP-BY-STEP SUMMARY

Joists

First floor joists, for a two-storey extension, will be built in by the bricklayers, but flat-roof timbers are the area of the carpenter. A chipboard decking is common to most flat roofs, but it must be kept dry at all times. Wet chipboard deteriorates very quickly.

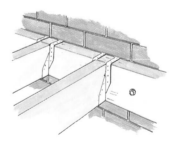

Cavity Trays

When the decking is on, and before the felt roofer arrives, a cavity tray must be installed where required. An angle grinder can be hired for this purpose, but great care must be taken to avoid injuries. Always wear goggles when using this kind of equipment. This may be a good time to add the flashing ready for the felt roofer to complete.

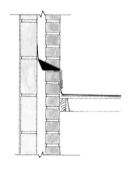

Herringbone

When the joists are in place, a row of her-ringbone struts can be fitted along a central line to toughen up the structure and reduce movement. Where the room is long and wide, two rows of struts may be required. The struts can be made from timber or ordered from your supplier.

Scaffold

The bricklayers will want scaffolding raised for a two-storey extension, or where gable end walls are being built. Keep in touch with the scaffolders to make sure that time is not lost.

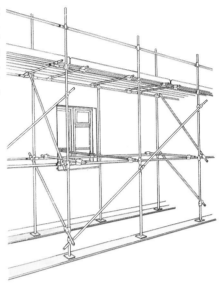

Felt Roofer

The felt roofer will quickly get your extension 'in the dry' as they call it. An upstand along the wall with the cavity tray will be left ready for a flashing to be added. If the flashing is already in place ask the roofers to dress it down over their felt upstand. Avoid walking on flat roofs unless it is absolutely unavoidable. The sharp granite chippings will easily puncture the surface and could lead to expensive repairs.

> **GUARANTEE**
> Some felt roof companies offer a guarantee for their work. If it is a reputable company, with many years trading, this may be a good offer to accept.

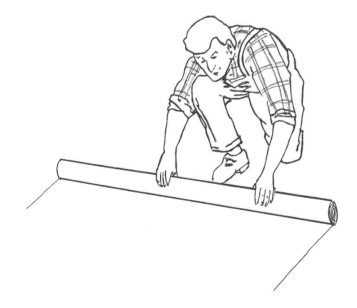

PHASE 4 – ROOFING 2 (PITCHED ROOFS)

Roof coverings have changed very little during the last 1,000 years. As soon as man left his cave and lived in the portable structures commonly still seen in use in some parts of Africa, simple roof coverings became essential. A sloping roof made from branches covered with dead plants and grasses progressed into the thatched roof we know and love today. Handmade clay tiles, dating as far back as the thirteenth century, were used to cover roofs in areas where suitable clay was found, eventually replacing the thatched roof as the number one roof covering. Clay tiles, now machine-made, can be seen in numerous shapes, sizes and colours throughout the country.

The roof structure, flat or pitched, has also progressed with a mass of regulations to ensure that it is strong enough to carry the load imposed upon it, insulated enough to assist energy conservation and ventilated enough to prevent condensation damage. Add to this the roof's appearance and how it compares with the existing building and you will realize how multi-functional a roof is now designed to be. The choice of roof shape and covering for your own extension may simply be a duplication of the existing roof, where even the slope will be at the same angle or, should this not be the case, perhaps on a shallow pitched roof where duplication is not possible, then a complementary alternative will have to be found.

PITCHED ROOF CONSTRUCTION

A roof is classified as a pitched roof when the pitch or slope exceeds ten degrees, though acceptable roof coverings for domestic extensions dictate that the slope is more likely to be in excess of 17½ degrees.

A pitched roof is a construction configured with rafters and ceiling joists where lean-to roofs and close-coupled roofs, and the less popular mono-pitched roof, will cover the majority of choices. There are two recommended ways of building a pitched roof.

The first and most common construction for home extensions, is a 'cut' roof. This is a pitched roof formed by the carpenter, on-site, from timbers cut to length to form rafters and joists. The timber dimensions will be specified on your plans but the timber lengths will not be, so leave the ordering to the carpenter who will make allowances for angles and overhangs.

An alternative to the 'cut' roof is the factory-made truss rafter. A truss rafter is a machine-made roof member of rafters and ceiling joists, it is structurally designed and made to suit your particular, individual requirements. Truss rafters are often a lot easier to install than a 'cut' roof, reducing labour costs by quite a lot. However, the material costs can be significantly higher for smaller projects like home extensions where only a few trusses are involved. Even so you may want to weigh up the total cost of a 'cut' roof, including labour, in order to compare it with the cost of a truss roof, including labour. Your local supplier will provide you with a quotation for the trusses and the carpenter will quote you for the labour.

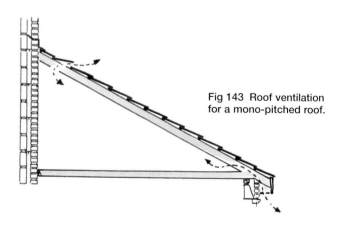

Fig 143 Roof ventilation for a mono-pitched roof.

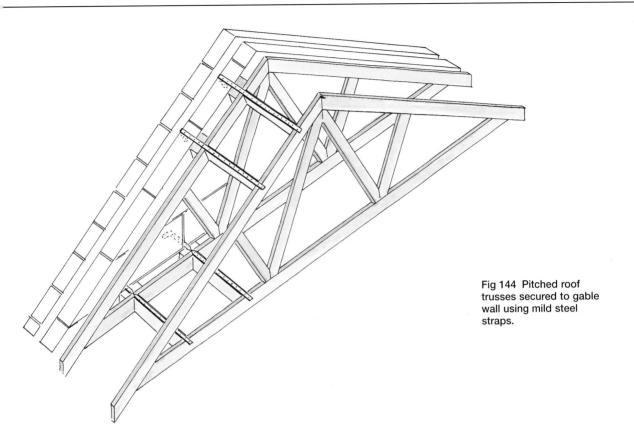

Fig 144 Pitched roof trusses secured to gable wall using mild steel straps.

Whichever roof structure you choose, when the timbers or the trusses arrive on-site they must be stored on a flat, level surface in a dry place until they are required for use.

LEAN-TO ROOF

A lean-to roof is the most common pitched roof style. It has a single slope, or pitch, designed to direct rainwater away from the existing building and into the guttering. The rafters on which the roof covering will be fitted will be secured to wall plates attached to both the existing building wall and the new building wall. The rafters may then require additional support, mid-span, depending upon the distance between the walls and the weight of the roof covering.

A lean-to roof can be built with a slope as low as 17½ degrees providing it is covered with large interlocking concrete tiles manufactured for this purpose.

CLOSE-COUPLED ROOFS

A close-coupled roof has a double pitch where two rafters are fixed to a ridge board at one end and to the wall plates, on opposite walls, at the other. A ceiling joist spans from wall plate to wall plate and is secured to the rafters to prevent them from spreading. Larger variations include purlins mid-way along the rafters, to increase the span without increasing the timber size, with timber struts where possible, directing the roof load down to a load bearing point. Binders and hangers will provide increased support to ceiling joists.

HIPPED ROOF CONSTRUCTION

A hipped roof is where the roof is continuous around the building, as opposed to stopping at a gable end. Hip boards are attached to the wall plates at one end and the ridge board at the other. A wall plate

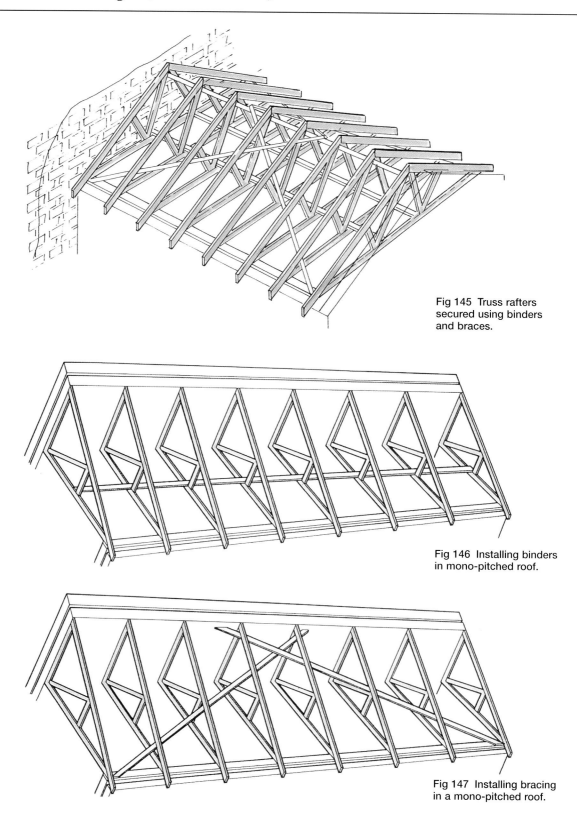

Fig 145 Truss rafters secured using binders and braces.

Fig 146 Installing binders in mono-pitched roof.

Fig 147 Installing bracing in a mono-pitched roof.

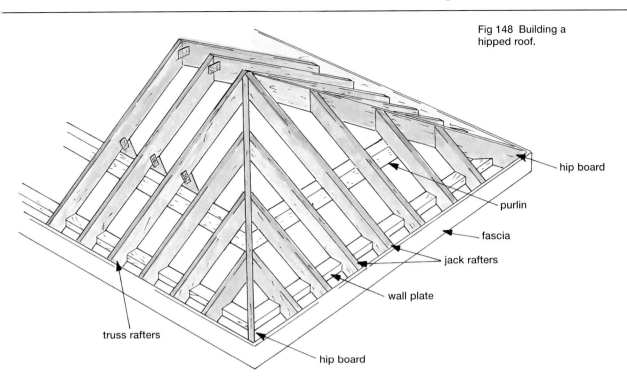

Fig 148 Building a hipped roof.

hip board

purlin

fascia

jack rafters

wall plate

truss rafters

hip board

will be attached to all the external walls of the extension off which the hipped roof will be built and braced across the corners to counter against spread. A hipped roof is far more complicated to construct than a close-coupled roof, with a large quantity of angles to cut. The smaller rafters between hip boards and ridge board at one end and valley boards and ridge board at the other end are called jack rafters. Covering a hipped roof also requires a greater degree of skill from the roof tiler with hip tiles and valley tiles adding to the need for accuracy during installation.

VALLEYS

Where a new pitched roof intersects an existing pitched roof at right angles, and part of the new roof is resting on the existing rafters, a valley will be formed. Valley boards, or layer boards, will be laid onto the existing rafters for the new roof to be built off. The existing roof timbers may need additional support to carry the extra load imposed upon it and structural calculations may be required to prove the support is sufficient. Where valleys are formed, weather proofing is

VALLEYS
When the roof timbers are in place the valleys can be added. Pre-formed or constructed from lead sheeting, they must be added before the roof tiling starts. The roof tiler's price will including fixing valley tiles but not necessarily valley linings. Before work commences make sure you understand what is, and is not, included in the price.

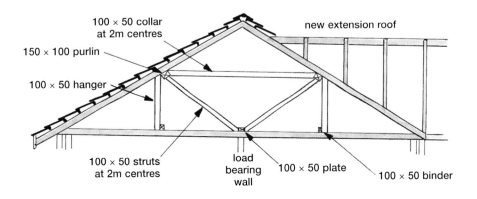

100 × 50 collar at 2m centres

new extension roof

150 × 100 purlin

100 × 50 hanger

100 × 50 struts at 2m centres

load bearing wall

100 × 50 plate

100 × 50 binder

Fig 149 Typical example of adding strength to a 'built on site' roof.

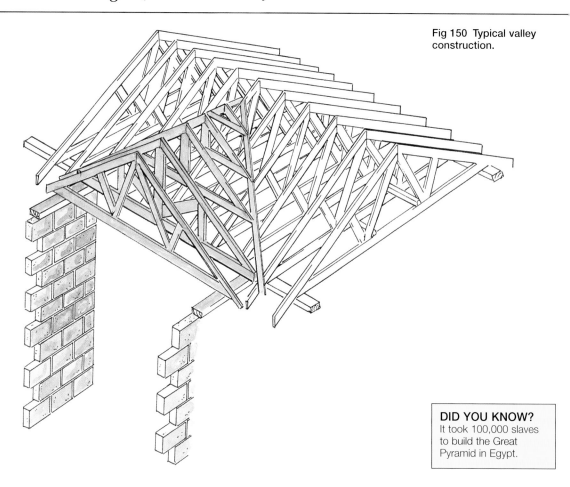

Fig 150 Typical valley construction.

DID YOU KNOW?
It took 100,000 slaves to build the Great Pyramid in Egypt.

important as water enters the valleys from more than one slope. This open gutter can be lined with lead, copper or fibreglass, or purpose-made valley tiles can be used as valley liners.

GABLE LADDER

Where the roof overlaps a gable end wall, a support for the overhanging tiles will be needed. If the overhang is small, then a rafter on the outside of the wall structure may suffice; however, when the overhang exceeds 150mm then a gable ladder may be needed. Known as a 'ladder' only because of its appearance, it will be constructed as shown in Fig 151 using the final internal rafter attached to a similar rafter, set externally using timber noggins. The barge boards and soffit can then be fixed to provide the decorative covering.

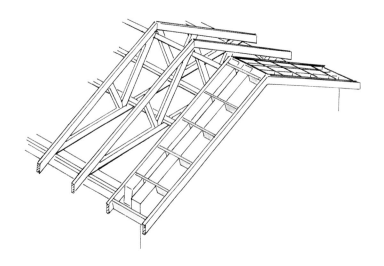

Fig 151 A gable ladder.

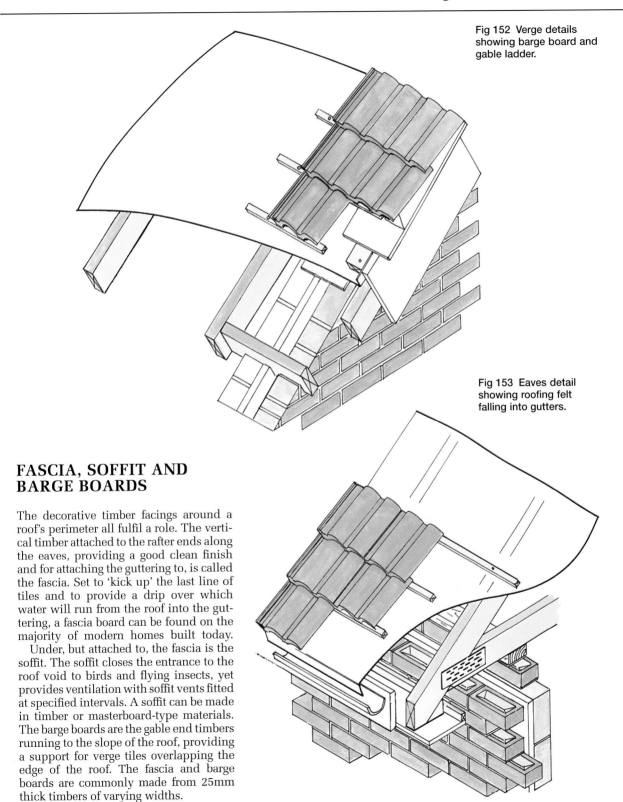

Fig 152 Verge details showing barge board and gable ladder.

Fig 153 Eaves detail showing roofing felt falling into gutters.

FASCIA, SOFFIT AND BARGE BOARDS

The decorative timber facings around a roof's perimeter all fulfil a role. The vertical timber attached to the rafter ends along the eaves, providing a good clean finish and for attaching the guttering to, is called the fascia. Set to 'kick up' the last line of tiles and to provide a drip over which water will run from the roof into the guttering, a fascia board can be found on the majority of modern homes built today.

Under, but attached to, the fascia is the soffit. The soffit closes the entrance to the roof void to birds and flying insects, yet provides ventilation with soffit vents fitted at specified intervals. A soffit can be made in timber or masterboard-type materials. The barge boards are the gable end timbers running to the slope of the roof, providing a support for verge tiles overlapping the edge of the roof. The fascia and barge boards are commonly made from 25mm thick timbers of varying widths.

Phase 4 – Roofing 2 (Pitched Roofs)

FELT AND BATTEN

Roofing or sarking felt will provide a secondary protection to the roof area from wind-driven rain and snow. It should be non-tearable and will be suitable for this purpose. An overlap of at least 100mm must be allowed where one layer joins another and secure joints around openings, soil vent pipes for example, must be formed. The felt will be secured at eaves' level and additional strips will be added to hip and valley areas.

Softwood roofing battens, treated with a preservative, will secure the roofing felt and, where battens are joined, the smallest battens should span at least three rafters. Batten sizes are generally 32mm × 19mm for rafters up to 450mm centres and 32mm × 25mm for rafters up to 600mm centres. The nails or fixings used must be appropriate for this purpose.

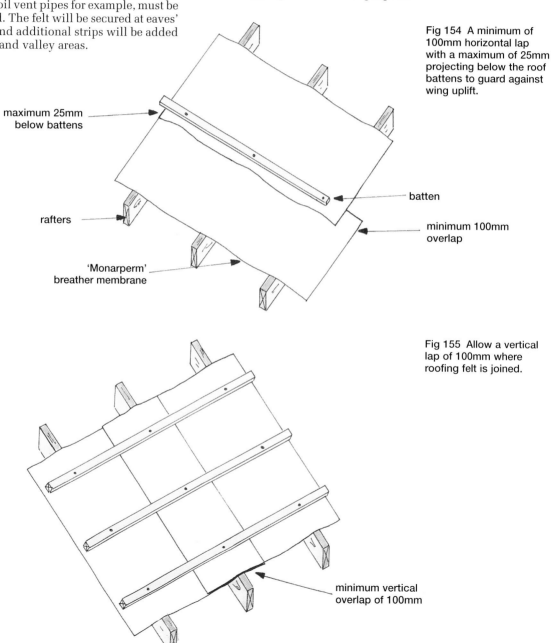

maximum 25mm below battens

rafters

'Monarperm' breather membrane

batten

minimum 100mm overlap

Fig 154 A minimum of 100mm horizontal lap with a maximum of 25mm projecting below the roof battens to guard against wing uplift.

Fig 155 Allow a vertical lap of 100mm where roofing felt is joined.

minimum vertical overlap of 100mm

TILES

Pantiles, interlocking tiles and plain tiles form the majority of roof coverings in a huge variety of styles and colours. The plain tile, in clay and concrete, is slightly cambered to assist water discharge off the roof and into the gutters.

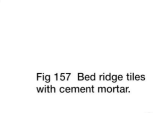

TILE COLOUR
It is not always possible to get exact colour matches with new tiles. If you are trying to match very old tiles there are many reclamation yards where a match may be found. It may take some time but it will make all the difference when you stand back and look at your completed works.

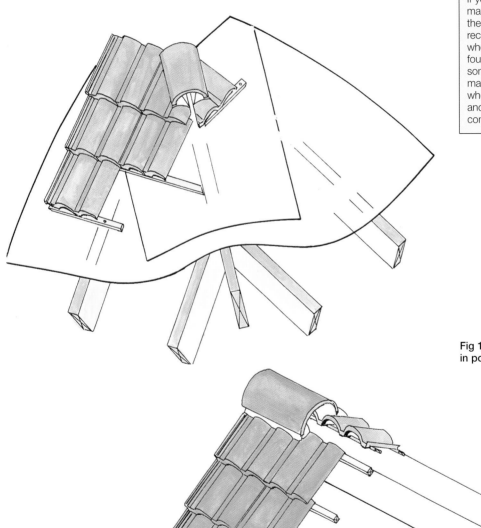

Fig 156 Hip tiles bedded in position.

Fig 157 Bed ridge tiles with cement mortar.

93

Phase 4 – Roofing 2 (Pitched Roofs)

While tiles and half tiles can be used at verges to provide staggered joints, eaves tiles should be used along the eaves to support the fall into the guttering and pre-formed valley tiles should be used to form the valleys. To complete the roof covering where one slope meets another, special tiles called ridge tiles and hip tiles, available in all shapes and sizes, will be fitted.

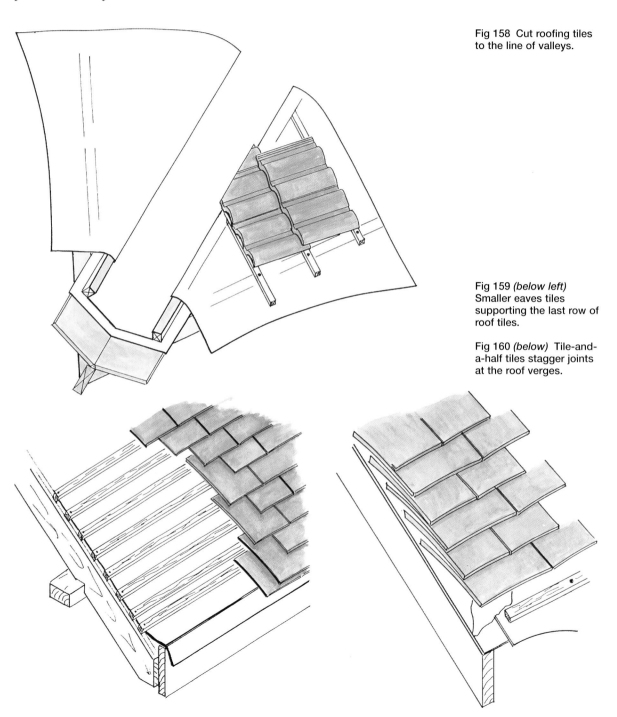

Fig 158 Cut roofing tiles to the line of valleys.

Fig 159 (below left) Smaller eaves tiles supporting the last row of roof tiles.

Fig 160 (below) Tile-and-a-half tiles stagger joints at the roof verges.

SLATES

Slates, natural and manufactured, are not as common as the clay and concrete tiles but are still in use in most areas. The slate is secured by two, usually copper, nails with an overlap covering the nails. The first row of slates is then covered by the second row with a half-slate stagger, providing a half-bond to reduce wind damage to what is quite a light roof and to ensure watertightness. Ridges and hips are often seen in clay to provide a clear outline to the roof.

CROSS VENTILATION

Roof voids provide the perfect opportunity for condensation to cause problems in well insulated and energy efficient modern buildings. For this reason Building Regulations stipulate that sufficient ventilation of the roof void must be provided. The normal method of roof ventilation is to provide continuous eaves ventilation, equivalent to a 10mm gap along the total eaves length, along opposite sides of the building to provide an unobstructed airflow above the insulated ceiling. Roof and rafter ventilators are available to assist airflow over the insulation quilt laid at the eaves. Air bricks built into gable walls at a high level will provide alternatives for lean-to roofs, or roofs where eaves ventilation in opposite walls is not possible. Special channels and soffit vents can be built into roofs at eaves' level to support continuous airflow over the insulation without blocking roof ventilation.

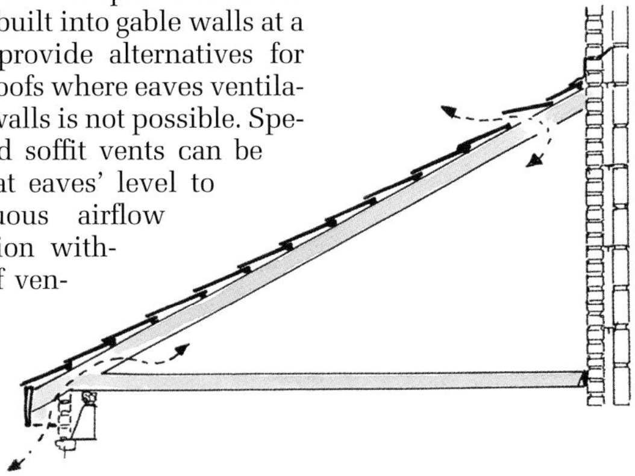

Fig 161 Roof ventilation for a mono-pitched roof.

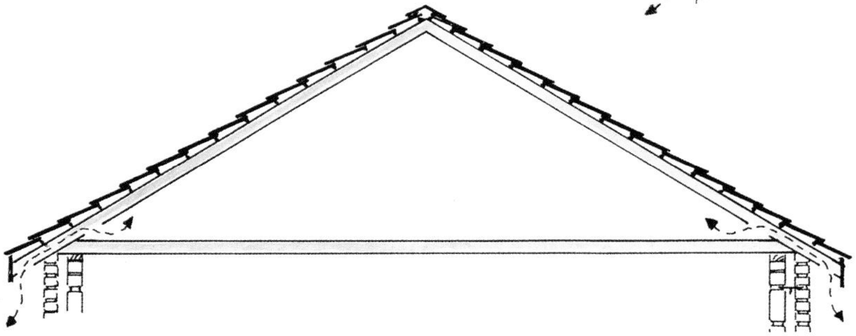

Fig 162 Eaves ventilation for a new pitched roof.

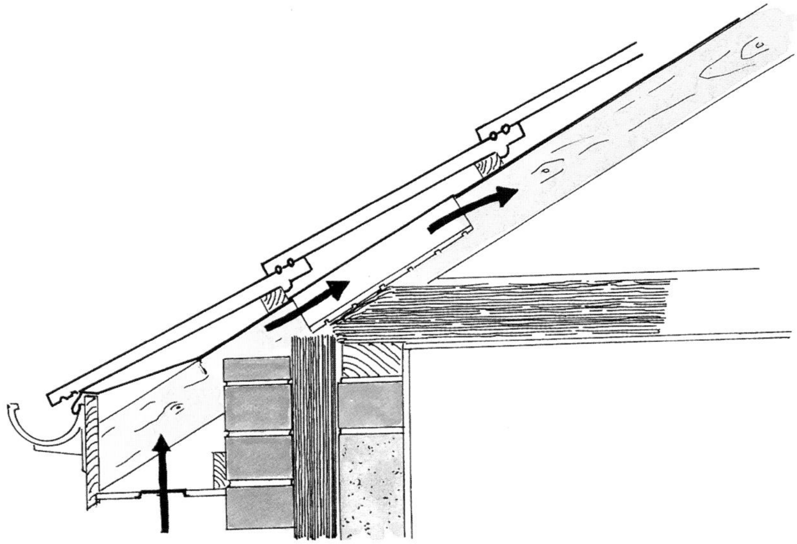

Fig 163 Install eaves ventilators in soffit boards.

Phase 4 – Roofing 2 (Pitched Roofs)

GUTTERING

With the scaffolding still *in situ*, and when the roof covering is completed, the rainwater guttering can be fitted. There are a wide variety of styles and colours in cast iron, plastic and aluminium to choose from, but it is likely that you will use a gutter identical to the existing property. Gutter brackets should be set at approximately 900mm centres, under standard domestic conditions, and downpipe brackets set at 2m centres. These measurements are for guidance purposes only. It is worth noting that this may be a good time to decorate the fascia boards and other timbers off the scaffolding and before the gutters are fitted. Good planning at this stage will save both time and money later on.

CARPENTRY

Building a pitched roof, on-site, is not a job for the inexperienced. Always use the most experienced tradesmen, as any extra cost is likely to be offset by the quality of the work. Snug fitting joints at ridge and eaves, bird's mouth fittings at wall-plate level, not to mention the additional work involved with a hipped roof, will aid the strength and stability of the roof structure.

At this time there may be a lot of work for the carpenter to carry on with after the roof structure is complete including

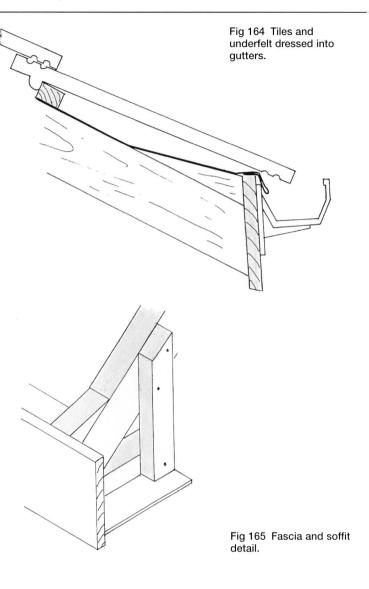

Fig 164 Tiles and underfelt dressed into gutters.

Fig 165 Fascia and soffit detail.

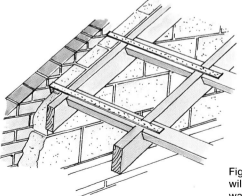

Fig 166 Restraint straps will secure the roof to the walls.

TIMBER PRESERVATION

Timber built into permanent structures will need to be treated against rot and insect damage. An advantage with using chipboard flooring sheets is that they are generally unaffected by rot or beetle attack which will, if permitted, cause terrible damage to untreated timbers.

New floor boards should be treated before they are installed, using a suitable timber preservative. There are two common types of preservative which can be brushed on, sprayed on or dipped to provide maximum protection. One is water-based, is odourless and soaks into the timbers, the other is white spirit-based and is commonly used by timber treatment companies. In both cases the water or white spirit evaporates leaving the chemicals to fight attack.

flooring, partitioning and so on. Try to get a price for *all* the carpentry work, or separate prices for different phases, and be sure to pay only for work that has been completed.

QUOTATIONS

The carpenter will quote to do the work on one of two schemes. The first, and most common, is 'price work'. Working from the drawings, a price to complete the work will be submitted for your acceptance. It may be broken down into phases of say the roof, the flooring and finishings, or it may be a total, completed price. The work will not include bedding on the wall plates, the bricklayer is responsible for this, but will include, unless otherwise specified, stripping the tiles where a new roof abuts an existing roof, for valleys to be formed and so on, and fitting fascias and soffits. The carpenter may also be the best person to fit the guttering, though this will not generally be included in the price given.

Secondly, the carpenter can work on a 'day rate' basis. This is not commonly used for standard home extension work but more applicable to difficult conversion works. On a day rate basis the carpenter will agree a price per day with you, and the work to be carried out on a day rate basis, he will then expect to be paid at the end of each week.

Quotes will also be required from the roof tiler for adding the roof covering. As with the carpenter there are two common methods of employment, but unlike the carpenter they are 'Supply and fix' and 'Labour only'. Larger roof tiling businesses may well offer supply and fix rates. This method may seem expensive but it will reduce the risk of over-ordering and wastage. It is not unusual in these instances for a small deposit to be requested for the materials. If you are quite happy to supply all the materials then a labour only price is what you want. The price will include laying all felt, battens and tiles, forming valleys and bedding verge and ridge tiles. Unless specified in the price, it is unlikely to include installing lead valleys or fixing lead flashings. For small home extensions this work should be carried out on a 'pricework' basis of one payment when the roof is complete.

TRADESMEN

This phase of work is likely to involve the bricklayer, carpenter and roof tiler.

- Plant hire: scaffold to remain until roof is completed.
- The 'labour only' roof tiler may require an angle grinder.

Single-Storey Extension

- The bricklayer: will be required to build up gable ends after the carpenter has set out the rafters.
- The carpenter: will build the roof to completion.
- The roof tilers: will felt, batten and tile the roof to completion.
- Labour: for fixing lead flashings and valleys. Fitting guttering.

Two-Storey Extension

- The bricklayer: will be required to build up gable ends after the carpenter has set out the rafters.
- The carpenter: will build the roof to completion.
- The roof tilers: will felt, batten and tile the roof to completion.
- Labour: for fixing lead flashings and valleys. Fitting guttering

STEP-BY-STEP SUMMARY

Ridge Board

To set out the roof timbers the ridge board needs to be erected, attached at one end to the existing building and the other to the support rafters. Always ensure these structural timbers are level and totally secure. From the ridge board both hips and valleys will be constructed.

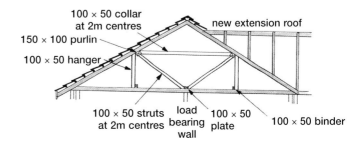

Phase 4 – Roofing 2 (Pitched Roofs)

Gable Ends

When the ridge board is in place the gable end rafters will support one end. They will also provide a template for the bricklayer to build the gable end wall. The gable ladder will also be built in at this stage.

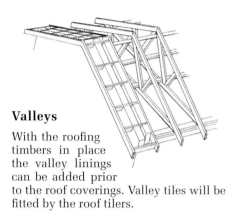

Valleys

With the roofing timbers in place the valley linings can be added prior to the roof coverings. Valley tiles will be fitted by the roof tilers.

Strip Roof

The roof will be stripped for this purpose. Remove tiles but leave felt and battens in place. Set aside the tiles removed for use at a later date. Try to keep breakages to a minimum. Re-using existing tiles helps the new roof blend in with the old.

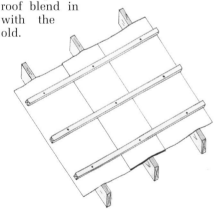

Roof Timbers

The carpenter will complete the roof timbers, infilling with rafters and ceiling joists. The gable ladder fitted before the bricklayer has built up the gable wall may require barge boards, fascia and soffits to be fitted.

Roof Tiling

Felt, battens and tiles can be added to the roof, finally getting the building 'in the dry'. Any existing roof tiles to be re-used will be best used where the roof is most visible, usually the front elevation, to hide any joins between new and old. The tiler will cut up the valleys and secure the verge tiles as required.

Guttering

With the scaffolding in place and the roof completed, the guttering can be fitted and connected to the water surface drains.

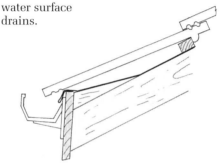

Off Hire

When the roof tilers have finished, and if it is no longer required, the scaffolding can be removed and returned off hire.

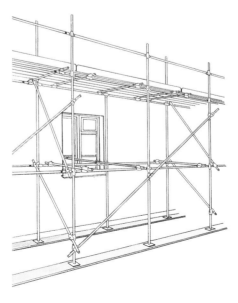

EXISTING TILES
Set aside any existing tiles to be re-used when the roof is covered. Older tiles are usually added to the facing side to hide an obvious join between new and old, especially where the new tiles are not exactly the same colour as the old ones.

PHASE 5 – BUILDING (INTERNAL WORKS)

Now, with the building watertight and the bulky structural building works complete, it is time to take a good look around the site and make preparations for the remaining outside work to be finished off before you begin the internal structural works. Timetables and flow charts should be updated, even rewritten, and a financial overview prepared. A list of materials already on-site can be compared to a list of materials required, with any surplus stored away or disposed of. The first phase of tradesmen will have finished, or will be preparing to continue with the internal works, and the tradesmen building the second phase will be keen to know if their schedules need to be changed. The weather should not now affect any of your arrangements leaving tradesmen, materials and money as the the only three areas where problems might arise.

It is not uncommon for this stage of the work to take expenditure to a new high. The materials required are often not large and bulky, but small and costly and include such items as double glazing and bathroom suites. A line should be drawn under the existing accounts position at this stage so that you can calculate how your expenditure

has gone and how it affects the work still to be carried out.

If you are building a single-storey extension, the floor and the partition walls will need to be ready before the plumber and electrician can carry out their first fixing. If the extension is on two storeys then the first-floor decking, and possibly a staircase, must be in place.

More importantly at this stage, from the outside at least, the extension will look finished so there is no need to leave the garden looking like a building site. The surplus materials will have been removed or relocated in readiness for new materials to arrive, allowing room for the site to be tidied up so that the garden can be reclaimed and replanted. All the other materials on-site should be stored away and the scaffolding returned off hire. The drains should be covered and the pipes set to receive downpipes or other drainage fittings and, if the soakaway is full, it too can be covered over. If it isn't full and there is likely to be more rubble building waste then leave it until it is.

To a builder this is a very exciting time. A time when all the preparations and planning have come to fruition and, like the first sign of spring to a gardener, the hard work is ready to bear fruit.

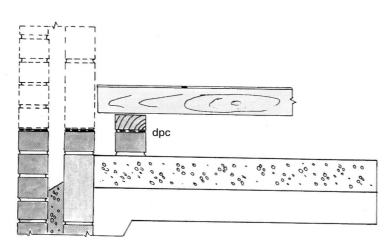

Fig 167 Suspended timber floor, on sleeper wall, at ground level.

TIMBER FLOOR JOISTS

A timber floor constructed at ground level is called a Suspended floor and a timber floor built at first-floor level is called a Single or Upper floor.

A 'suspended' floor is a perfect explanation for floors built at ground level. The floor timbers are suspended above the concrete oversite resting on brick-built

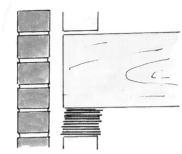

Fig 168 Do not use excessive packing when installing floor joists.

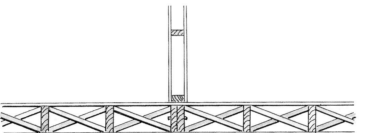

Fig 169 Double joists can be used to support partition walls.

'sleeper' walls. The sleeper walls are a single brick width and will be at least one brick high with a wall plate bedded on to receive the floor joists. Where the sleeper walls exceed one brick in height the walls used to be built to produce a honeycomb effect to allow for ventilation. More commonly seen today are airbricks built in at approximately 650mm centres, allowing free passage of air under the flooring assisted by the airbricks already built into the external, extension walls. Sleeper walls are built up to damp proof course level with no flooring timbers built into the wall below ground level.

Suspended floor joists are usually 100mm × 50mm on a 100mm × 50mm wall plate with underfloor insulation fitted between the joists.

A Single or Upper floor in a domestic extension will normally be formed using 200mm × 50mm timber joists set at up to 600mm centres with chipboard flooring or floorboards nailed on. Variations in joist size and centre will be clearly specified on your plans and may vary relative to the span between supporting walls. (For First Floor joists *see* Chapter 7.)

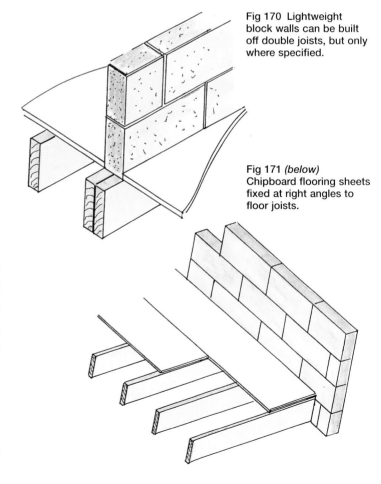

Fig 170 Lightweight block walls can be built off double joists, but only where specified.

Fig 171 *(below)* Chipboard flooring sheets fixed at right angles to floor joists.

TIMBER FLOORING

The floor covering, where timber joists are installed, will be a choice of either timber floorboards or sheet coverings in chipboard and plywood. Softwood floorboards are tongued and grooved and available in a variety of widths up to 150mm. The boards will be secured at right angles to the joists, using floor nails called brads, and any joins must be formed directly over a floor joist. To avoid the risk of dampness, for floors at ground level, leave a small gap around the perimeter of the floor. These small gaps will be concealed when the walls are plastered and the skirting board is fitted, whilst still allowing air to circulate.

As an alternative to floorboards, chipboard sheet floor covering can cover a large floor area very quickly and, for this reason, it is very widely used by the domestic house building industry. The tongued and grooved sheets can be ordered in 2,440mm × 600mm sheets, and from 18mm to 22mm thick. The sheets must be of flooring grade chipboard and should be moisture resistant.

OTHER FLOORS

A 'floating' floor will consist of a layer of insulation, perhaps polystyrene or mineral fibre sheets, on which flooring sheets, usually chipboard, or a sand and cement screed, are laid. A damp proof membrane will be installed to reduce the risk of rising damp and will form a continuous barrier with the existing wall damp proof course.

FLOOR INSULATION

Building Regulations are particularly stringent when it comes to preventing heat loss through solid and timber floors. To comply with them, and to meet the minimum requirements, a layer of insulation must be installed below floor level. The approved building plans will include specifications for the floor construction and insulation that is required. Choices will vary from insulation slabs laid directly onto concrete oversite for a floating floor to insulation secured between floor joists for a suspended floor. Your building plans will also clearly specify the type and grade of insulation that is required along with the installation methods.

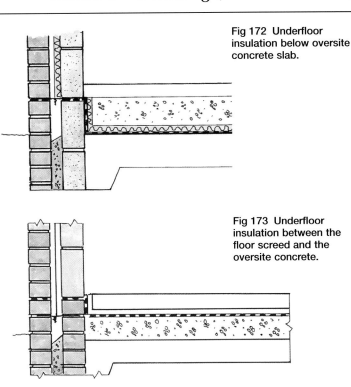

Fig 172 Underfloor insulation below oversite concrete slab.

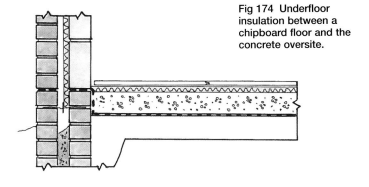

Fig 173 Underfloor insulation between the floor screed and the oversite concrete.

Fig 174 Underfloor insulation between a chipboard floor and the concrete oversite.

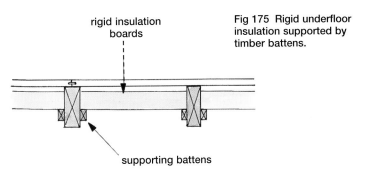

rigid insulation boards

supporting battens

Fig 175 Rigid underfloor insulation supported by timber battens.

Phase 5 – Building (Internal Works)

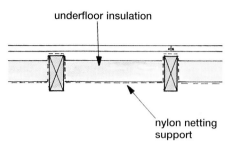

Fig 176 Underfloor insulation supported by netting.

FLAT ROOF INSULATION

Thermal insulation has become a byword with energy conservationists where any

require one, but it is a fairly unusual addition.

For staircases, Building Regulations are drafted to make sure that safe passage from one floor to another is provided. The type of staircase you would like, which can vary from a straight flight to a spiral, the floor space for the staircase and sufficient headroom going up the stairs, all have to be considered as these are significant and important factors when trying to comply with Building Regulations.

Here are a few examples of the staircase terminology used in the trade:

- The 'string' is the outer board providing support for the steps.
- The 'tread' is the board you put your foot on when using the stairs.

Fig 177 At least 50mm free air passage must be left above roof insulation.

Fig 178 Headroom measurements taken from the stair string and the stair tread.

domestic structure is concerned, and that includes the roof structure. A flat roof, in particular, is a complex design, incorporating energy conservation while also coping with ventilation to disperse moisture produced inside the building and providing protection from water produced outside the building.

A 'cold' flat roof, where the insulation is installed between the roof decking and a vapour barrier at ceiling level, and a 'warm' flat roof, where the insulation is built in between a vapour barrier laid on top of the decking and the waterproof roof layer, will both need to meet the same insulation requirements. Roof insulation quilt must be kept dry at all times if it is to work properly.

STAIRCASES

Not all home extensions will include the installation of a new staircase. Perhaps a new granny annex, where the upper floor is separate from the existing upper floors, or a conversion of the loft area will

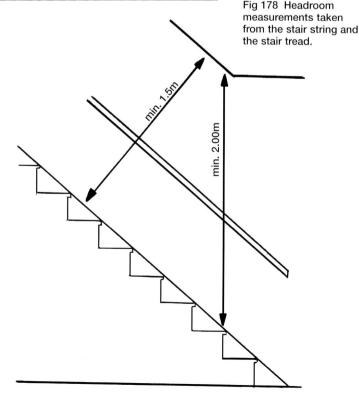

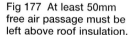

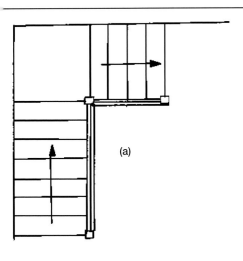

(a)

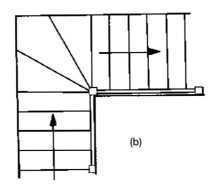

(b)

Fig 179 Staircase designs:
a. quarter landing;
b. kite winders;
c. straight flight staircase with a bullnose bottom step;
d. a half landing.

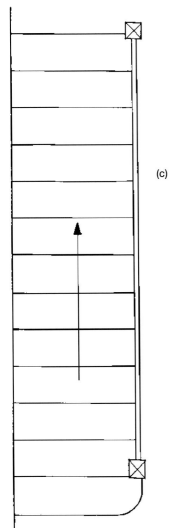

(c)

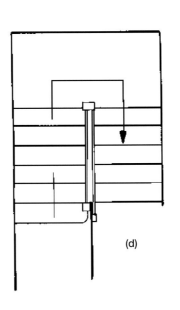

(d)

STAIRCASES

When the staircase flight is installed, cover and protect any surfaces which will be visible after the work is completed. Feature stairs, softwood or hardwood, can easily be damaged as the finishing tradesmen complete their work. Any damage from knocks, mortar, or plaster droppings may be extremely difficult to remove or disguise.

STAIRCASE DESIGN

Because of all the possibilities available, designing a staircase can be extremely demanding. Given enough headroom there is also the location and style to consider – where you want the staircase to start and end. Safety aspects must also be considered which include staircase width and handrail selection. Unless you have experience with staircase design, advice should always be sought before finalizing decisions.

Phase 5 – Building (Internal Works)

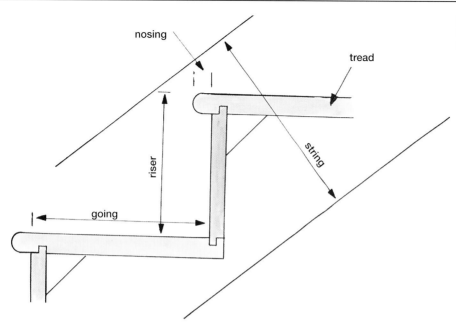

nosing

tread

riser

string

going

- The 'going' is a measurement of the tread taken from the nosing of one tread to the nosing of the next tread.
- The 'riser' is a measurement taken from the top of one tread to the top of the next tread.
- 'Newel posts' are the thick vertical posts, often at either end of the string.
- A 'Handrail' is the strip of wood fixed between the newel posts to give assistance to stair users.
- 'Balusters' are the small vertical posts supporting the handrail.
- The 'pitch' is the angle of the staircase taken at the lower floor level.

Designing a new staircase can be quite a daunting task so here are a few of the calculations to help you:

- The pitch of a staircase should not exceed 42 degrees.
- Each 'going' must be a minimum of 220mm and each 'rise' must be no more than 220mm.
- Clearance: A measurement taken vertically from every step must show a headroom clearance of 2m and a line drawn at 90 degrees from the pitch line should show a minimum clearance of 1.5m.

And finally: the relationship between the 'goings' and the 'risers' is the deciding factor which ensures that a staircase is neither too shallow nor too steep. A formula to calculate this is:

2 Risers + 1 Going = a sum between 550 and 700mm.
(Stair formula: 2R + 1G = 550mm to 700mm.)

DOOR LININGS

Unlike door frames, softwood door linings are for internal doors and consist of two Jambs and a head with loose door stops. The door stops can be positioned when the door is hung to produce an appropriate rebate for the door, ensuring a snug fit. You can select from a variety of sizes all relevant to the thickness of the partition walls the linings are to be built into. Frame ties should be used when fixing to block partition walls and screws or nails should be used when fixing to timber partitions.

If the linings are to be stained or varnished rather than painted, be sure to cover the frame before it is exposed to mortar or it will stain badly. Door linings can be ordered already assembled or they can be built on-site when required.

INTERNAL LOAD-BEARING WALLS

The majority of home extensions consist of more than one room, requiring an internal wall to be built. The wall will be 'load bearing' if it is included in the calculations for any floor or roof joists resting on it. This does not mean that when joists pass over a wall that it will be load bearing; the joists in question may be able to span from wall to wall without requiring support from the wall it passes over. A load-bearing wall will be shown as such on the plans, and will be built off a proper foundation excavated along with the other foundations and built up with the other walls, to damp course level and so on. The wall may only be a lightweight wall built from lightweight blocks, but if it is load bearing it will still need a foundation. Any openings for doors or windows or even hatchways in a load-bearing wall will require a lintel installed to Building Regulation requirements.

INTERNAL NON-LOAD-BEARING WALLS

Unlike load-bearing walls, where a foundation of sorts is always required, it is possible to build non-load-bearing walls directly on the oversite slab. The only provision is that the oversite slab must be able to carry the added weight of the new wall. If the new wall is to be built with lightweight blocks, bricks or concrete blocks, then it is likely that the oversite concrete slab will need to be strengthened along the line of the new wall.

> **DID YOU KNOW?**
> The purest form of gypsum is found near Paris. It is white and known commonly as Plaster of Paris.

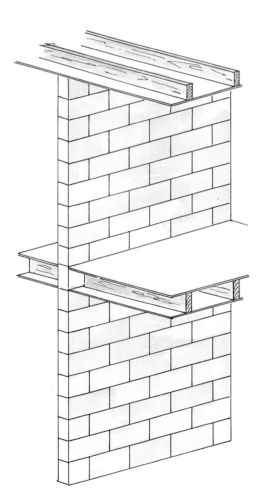
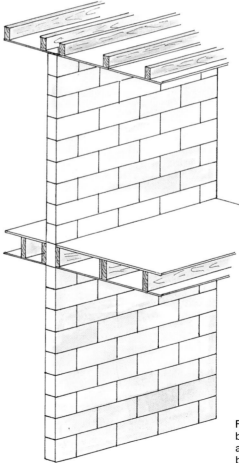

Fig 181 Load-bearing wall *(far left)* and non-load-bearing wall *(left)*.

Phase 5 – Building (Internal Works)

TRADESMEN

This phase of work is likely to involve the carpenter and the bricklayer.

- Plant hire: cement mixer.
- The bricklayer: will build internal block partition walls.
- The carpenter: will lay the floors, install the stairs and assemble the door linings.
- Labour: will install flat-roof insulation and tidy the site.

STEP-BY-STEP SUMMARY

Timber Flooring

Now that the building is 'in the dry', the ground floor and first-floor flooring can be laid. Make sure finished floor levels coincide with the existing floor levels before laying the floors.

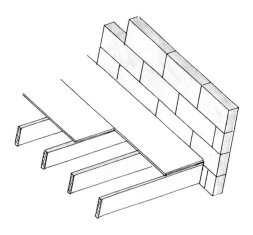

Staircase

If a staircase is to be installed, the flight can be secured to walls and joists prior to plastering. The handrails and newels can be added later if necessary.

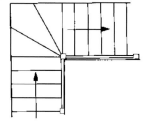

Block Partitions

The bricklayer may be required to build block partition walls and clear up any loose ends from his first visit.

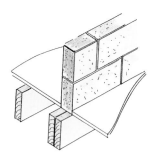

Breaking Through

When the windows are installed and the new building is secure, a section of wall will be removed to connect the old and the new together. This may involve installing lintels or RSJs. An angle grinder will provide a good, clean finish when removing sections of wall. For safety reasons use the goggles provided.

Soakaway

Waste rubble can be dumped into the soakaway or removed from the site if not required.

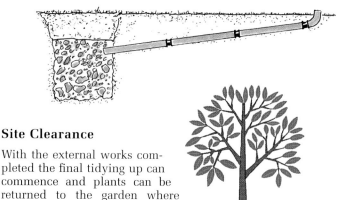

Site Clearance

With the external works completed the final tidying up can commence and plants can be returned to the garden where required.

load-bearing wall

adequate joist bearing

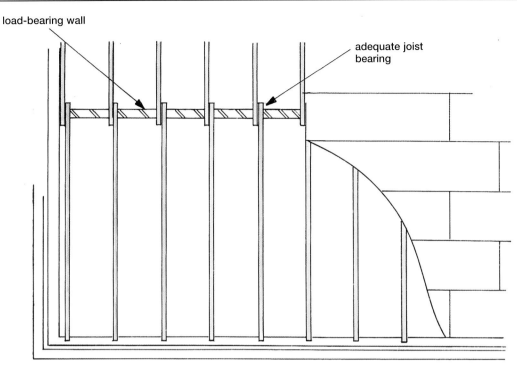

Fig 182 Joists resting on load-bearing wall *(above)* and joists trimmed around stairway opening *(below)*.

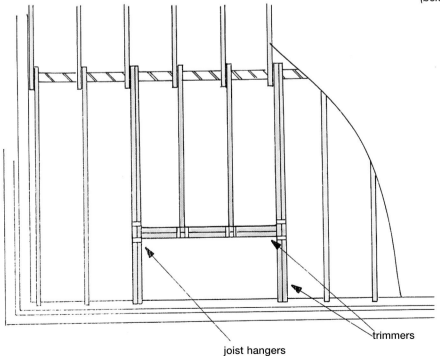

trimmers

joist hangers

FIRST FIXING

It is without doubt that the 1870s was an important decade for the British householder. The first raw-sewage treatment plants had just started to operate around our cities and Thomas Twyford patented the first ceramic flush toilet. At the same time a certain Thomas A. Edison was producing the first commercial filament lamp and designing the world's first Electricity Power Station, built in London in 1882.

One hundred and twenty years later and there is little change to these earlier designs. Naturally the process has improved and become more orderly, but the fact that the modern twentieth century home has entered the new millennium with filament bulbs and flush toilets connected to power stations and sewage treatment plants, respectively, bears full testament to these two great inventors.

The story of electricity as a source of energy for the home actually dates back to the time of Sir Humphrey Davy around the 1820s, but the story of plumbing has a far greater historical background. Taken from the Latin word for lead, *plumbum*, historical finds dating back to before the Romans show plumbing and drainage has a basis well ensconced in the history of domestic dwellings. Progress through the Middle Ages was particularly slow though, and it wasn't until the end of the nineteenth century and the inventions by Twyford and Edison that significant progress was made.

With the majority of major structural works now completed, it is time for the supplies to services, or first fixings, to be started. These include wires and boxes for electrical fittings and pipes for sanitary works that will be hidden in voids behind ceiling boards and plastered walls.

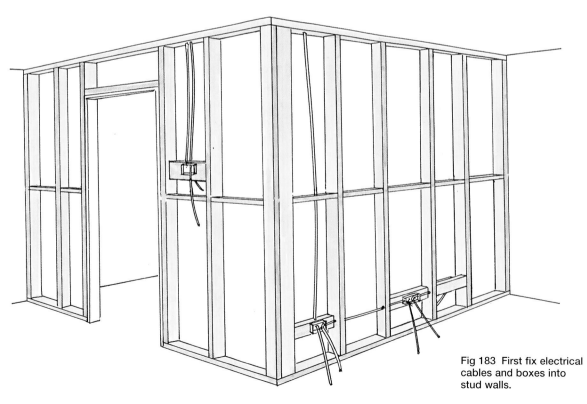

Fig 183 First fix electrical cables and boxes into stud walls.

FIRST FIX: ELECTRICS

Your building plans may or may not show electrical points and light fittings because they are not currently governed by Building Regulations. Points of good practice and guidance to ensure good installation have been laid out in a hefty tome called *Wiring Regulations* which is produced by the Institute of Electrical Engineers.

Domestic installations are supplied from a small local sub-station where the power is reduced to 240 volts single phase supply. Cabled into homes underground or overhead, the supply is metered before connection to a fuseboard where fuses, or circuit breakers, are a precaution against overloading and short circuiting. At the fuseboard the supply is separated into fused sections to serve individual facilities such as the cooker and the water heater. Power points and plug sockets will have at least two fused sections as will the lighting circuits.

Overloading can be the result of using an appliance connected to a lower current supply, while short circuiting can occur when the live wire touches either the neutral or the earth wires. In older properties the electricity system was originally earthed by connection to water pipes, never gas pipes, but in modern homes the earth is commonly through an earth terminal.

The demands placed upon the modern electricity supply are due to the number of accessories available to connect to it

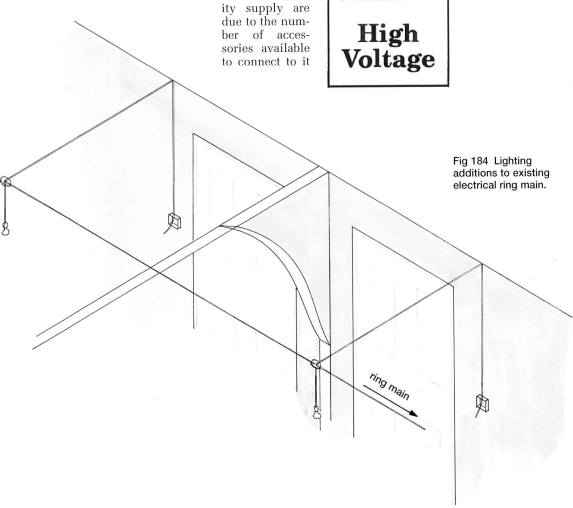

Fig 184 Lighting additions to existing electrical ring main.

ring main

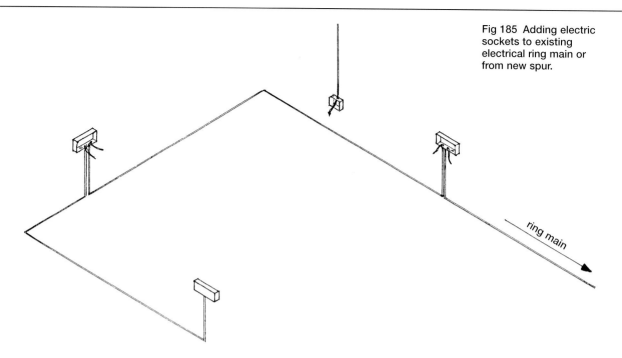

Fig 185 Adding electric sockets to existing electrical ring main or from new spur.

and require the services of qualified electricians *only* to make additions or alterations. The electrician will discuss your requirements with you and will need to inspect your incoming supply and fuseboard before a quotation can be given. Larger extensions can place quite a demand upon existing supplies so a new fuseboard may be required. Caution with electrical installations must be observed as this is not an area where penny-pinching can be considered.

Make sure the electrician is qualified. There are organizations, such as the National Inspection Council for Electrical Installation Contracting, that register qualified tradesmen, but make sure you get more than one quote. The electrician's first job will be to fix cables and boxes, known in the trade as a 'first fix', so that the ceilings can be boarded and the walls can be plastered. The cables will be used to carry the electricity supply to the new fittings and the boxes will be used to fix the fittings in place. When the cables and boxes are in position and the light fitting cables are secured, take a good look at the layout and double check that they meet with your requirements. If you want to add plug sockets or light fittings, now is the time, before the plastering starts.

And remember: electricity can *kill*.

FIRST FIX: PLUMBING

There are two areas where plumbing may be required for your extension. The first is the addition of radiators for central heating and the second is water supply for sinks, baths and toilets. The plumber must view your existing facilities before an accurate quotation can be given. Adding radiators to an existing system may put excessive demands upon the existing water heater, thus reducing its ability to heat the home properly. If this is the case a new water heater may be required. Also, if you are building a single-storey extension with a view to adding a second storey at a later date, this may be the best time to upgrade your existing water heater. Registration of approved installers to a known regulatory body such as CORGI will assist you when selecting your plumber because any work involving the gas supply must be carried out by a plumber with the appropriate registration.

The addition of a water supply for new facilities, such as a sink or a toilet, from the mains should be quite straightforward. The cold water supply to sinks should come directly from the mains supply, but the cold water supply for

PLUMBING
Make sure your existing central heating boiler is capable of coping with the additional load. Now may be a good time to upgrade the old boiler for a new one. Ask your plumber for advice.

ACROW PROPS
An adjustable support for walls, floors and roofs, required during demolition and when breaking through. Acrow Props can be hired from your local hire centre.

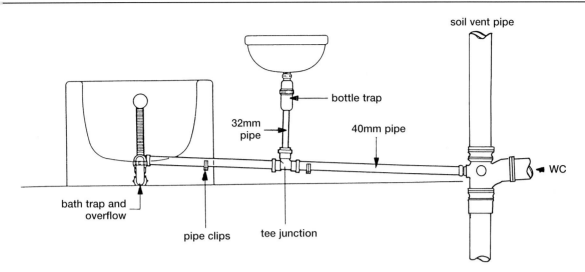

soil vent pipe

bottle trap

32mm pipe

40mm pipe

WC

bath trap and overflow

pipe clips

tee junction

Fig 186 Connecting appliances, services and utilities to a soil vent pipe.

baths and toilets can be from a cold water tank in the loft. Hot water will be supplied from the hot water tank.

The pipes to these facilities may be hidden behind ceilings or in floors. If this is the case, mark their location on your plans for future reference. If the pipes are to be visible, surface-mounted pipes, then fixing will be carried out after the walls have been completed and prior to decorating.

Plumbing is clearly one of the areas where the DIY enthusiast has been targeted and pampered. The materials used and the layout of many systems are designed to help where experience falls short. There are though, still areas where the enthusiast must seek professional assistance and installing a new gas boiler or a new hot water tank are just two such specialist areas. Adding radiators to an existing system, and even adding bathing and toilet facilities should hold few terrors for the keen DIY enthusiast. There is, however, one simple note of caution. If you lack experience jointing copper pipes you may find that concealing them in ceilings and under floors, before the system has been checked and found to be watertight, could easily result in expensive problems.

LOFT HATCH

Access to the ceiling void through a trapdoor, more applicable for a pitched roof than a flat one, must be positioned correctly. The position of the access hatch, apart from being in a location where a loft access ladder can be used, should be sensible without being over-prominent. The opening should be approximately 600mm square providing access to water tanks and various other services, including access to insulate the roof. A loft hatch opening can be framed with trimmers in a 'cut' rafter roof but should be sized to suit the spacing available in a truss rafter roof. Rebated 'linings' for the hatchway will be fitted with stops, similar to door linings and door stops, onto which the trapdoor will rest. The hatch lining should be set down from the ceiling joists, the thickness of the ceiling boards, so that a flush fit can be achieved.

For a flat roof, access to services such as electrics and plumbing may be necessary so a small, concealed area should be provided at the appropriate point.

CEILING BOARDS

The decorative finish you want to achieve for your ceilings will determine what materials you use. The most popular and very widely used in modern building construction are plasterboards, with a choice of Artex or plaster skim to provide the finish. Plasterboards are excellent for ceilings with an ivory-coloured finish side, suitable for direct

BREAKING THROUGH

Removing sections of wall to form an opening into the new building can be a major job. Make sure there are adequate supports for walls and floors where necessary.

When you do break through the existing wall into the new building, make sure dust sheets cover all the furniture and carpets. The mess can be excessive and very difficult to eradicate.

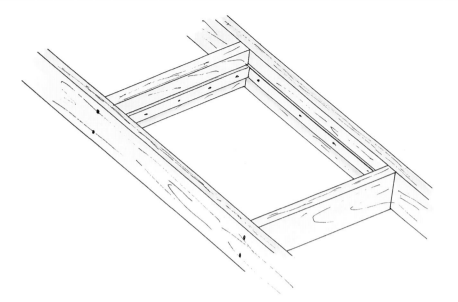

Fig 187 Add trimmers to roof joists to provide an access hatch.

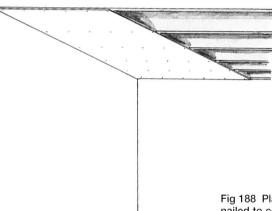

Fig 188 Plasterboards nailed to ceilings.

decoration, and a grey side suitable for plaster skimming. The boards are filled with a gypsum plaster core. When nailed in place, using galvanized nails, the joints can be taped or muslin jointed prior to decoration or plaster skimming. Plasterboards are available in a variety of sizes, 2400 × 1200mm is the most common, and thicknesses relevant to the span between the ceiling joists. Where joists have centres of up to 450mm then 9.55mm boards are acceptable and for joists with centres up to 600mm then 12.5mm boards are preferable.

FIRE RESISTANCE

Plasterboards provide an excellent fire resistance for domestic situations, protecting the main structural floor and roof timbers and reducing, or at least delaying, the spread of a fire. The plasterboards should be fixed lengthways at right angles to the joists and the ends must meet on a joist. The ceilings of a garage where the risk of fire is higher and where there is accommodation above, for example, will need to be double-tacked with ceiling boards with the joints staggered to provide a good seal. The joints should then be taped and the edges sealed along the walls to provide fire and fume protection for the upper floors.

LOFT INSULATION

To prevent up to 25 per cent heat loss from your home, the loft space will be insulated to comply with Approved Document L of Building Regulations. Available in either loose fill, or blanket quilt, the insulation should be a minimum of 100mm in depth covering the complete loft area.

Loose fill insulation is available as mineral fibre or vermiculite particles and can be poured between the joists, using a rake to level it off. Loose fill is particularly ideal for areas where access is limited or when poured into areas where sound insulation is required, for example, between floors and ceilings. The advantages of loose fill insulation are that

> **LOFT INSULATION**
> When you install fibreglass, or similar fibre insulation, in your roof void, be sure to wear protective gloves and a mask.

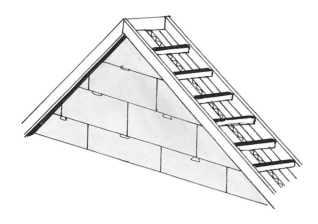

Fig 189 Continue cavity insulation up gable walls.

there is no cutting involved, it is non-combustible and does not leave gaps.

Quilt insulation is available in rolls and can be easily cut to suit the joist centres. Installation is also relatively simple, although gloves and masks should be worn at all times. The quilt can be laid tightly between the joists and tucked into the eaves, making sure to allow for ventilation at eaves' level. The advantages of quilt insulation are that it is likely to be cheaper than loose fill, it is also non-combustible, and is easy to cut.

Both loose fill and quilt insulation should not rot, should inhibit fungal, mould and bacteria growth and should be vermin proof.

Specifications for your extension insulation will be clearly indicated on your approved plans and any alterations to this must be approved with the Building Inspector prior to the work being carried out.

or softwood with matching timber panels, or rebated to receive glass panels, producing the required effect. A flush door will have an exterior grade plywood finish over a filled core, and will, occasionally, have openings for glass panels. With so many doors to choose from, and such a wide range of suppliers, any selection you make will be down to personal choice or simply to match existing doors.

Hanging a door into a new frame may require a little adjustment but hanging a door into an old frame is not recommended. If at all possible, install a new frame and a new door. And the door should suit the frame, for example a hardwood door would be complemented by a hardwood frame and a painted door would be complemented by a painted frame. The choice of door furniture is also extensive and will be based on personal choice.

VERMICULITE INSULATION

Loose fill vermiculite loft insulation is considered to fulfil the 'green' requirements so well publicized these days. A totally natural product, vermiculite is a loose fill loft insulation which is fibre-free, non-irritant and non-toxic. There is no need to wear gloves or masks, simply pour the sacks of expanded vermiculite between the rafters and rake level. There is no cutting or shaping involved and it is fireproof, energy efficient, cost effective and easy to install. (For more information see Useful Addresses at the back of this book.)

Fig 190 At least 50mm free air passage must be left above roof insulation.

EXTERNAL DOORS

Doors can generally be divided into two categories, panel doors and flush doors. The panel door is made from hardwood

MADE-TO-MEASURE WINDOWS

A made-to-measure window is a window made specifically for the opening avail-

113

First Fixing

Fig 191 Double-glazed units should be purchased from specialist suppliers.

able for it. There are several types of made-to-measure windows including UPVC, aluminium, hardwood and pine windows. When the walls are complete and the building is watertight the made-to-measure windows and doors can be fitted. The window board and door sills can then be added prior to plastering and floor screeding.

The window supplier will supply fixing instructions applicable to the window and these instructions should be followed carefully. If the carpenter is installing the windows, which would be preferable, then pass the installation instructions on.

GLAZING

Glass has been around for several centuries and became increasingly effective during the sixteenth century in Venice, when molten glass was spun into a flat dish shape then cut into rectangles. Clear plate glass however wasn't introduced until the nineteenth century and it wasn't until as recently as the 1950s that the Pilkington Brothers of England perfected float glass.

Float glass is the most common type of glass used for domestic fittings and comes in a variety of thicknesses from 4mm upwards. The thickness of glass required for your windows will depend very much upon the size of the opening and how far the opening is off the floor. For glass door panels, side panels and windows within

800mm of the floor level, the glass must be of a quality which, if broken on impact is unlikely to cause injury, such as toughened or laminated safety glass. The size of the opening will also govern the thickness of the glass required, while in bathrooms and toilets obscure glass should be fitted.

Before fixing glass into a wooden frame make sure that the rebate is cleaned, undercoated and painted with an oil-based paint. The rebate can then be puttied along the inner edge, prior to the glass being fitted, and held in place with small headless nails. The open rebate can then be puttied and finished off tidily. Glass in door panels will be held in place using glazing sealant and wooden glazing beads.

Fixing glass into a metal frames is a similar process; the rebates will be painted with an appropriate frame paint and the glass fitted as with a wooden frame, except that clips instead of nails will secure the glass.

THE GLAZIER

Unless you are a really keen DIYer and you find installing glass a fairly simple task, there are few other reasons for not using a glazier. The glazier will supply and fix the glass you require saving both time and money and eliminating wastage.

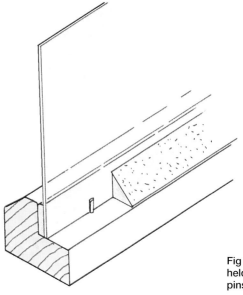

Fig 192 Glass will be held in place using metal pins and putty.

114

DOUBLE-GLAZED UNITS

There are two types of double-glazed window units: primary and secondary double glazing.

Primary double glazing is where two panes of glass are hermetically sealed together with a clear gap between them to form a 'sealed unit' which is then fitted into the rebated frame as a unit. Secondary double glazing is where the existing window has a separate and independent pane of glass fitted over it. It is important that, when ordering window frames, they have sufficient rebate to receive the double-glazed sealed units.

TRADESMEN

This is quite a busy phase with several tradesmen co-ordinating, including the carpenter, the electrician, the plumber and the glazier.

- The carpenter: will fix the loft hatch, hang the external doors and install the made-to-measure windows.
- The electrician: will supply first fix wires and boxes.
- The plumber: will supply and install water tanks and fix hidden pipes.
- The glazier: will supply and install glass where required.
- Labour: will tack the ceilings and install the insulation.

STEP-BY-STEP SUMMARY

Electrics

The electrician can now install the first fixing of cables and boxes, prior to the ceilings and walls being completed. Make sure there are enough sockets as it will be more expensive to add more at a later date.

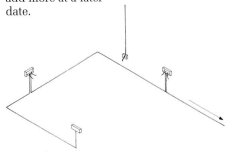

Plumbing

The plumber can install pipes and determine positions for new radiators. All the hidden pipes should be insulated and secured in place to avoid noises when the work is complete.

Windows

At this stage, with the external works completed, the specialist windows can be installed and the built-in windows can be glazed.

Doors

External doors can also be fitted to secure the building totally.

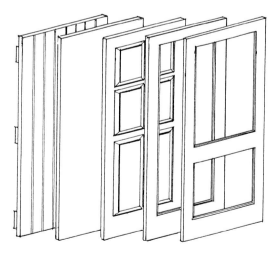

Fig 193 There will be a wide selection of doors to choose from.

Flat Roofs

Roof insulation will be installed in flat roofs prior to the ceilings being fitted.

First Fixing

Ceilings

For flat roof extensions the ceilings can be tacked, and for both single-storey and two-storey extensions, the ceilings can be tacked and the loft access hatch cut out. The roof areas where there is a pitched roof can be insulated.

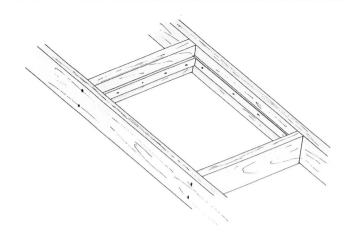

Stud Partitions

When the ceilings are in place the stud partitions can be added. Make sure any electricity cables and pipes are installed before they are covered with plasterboard.

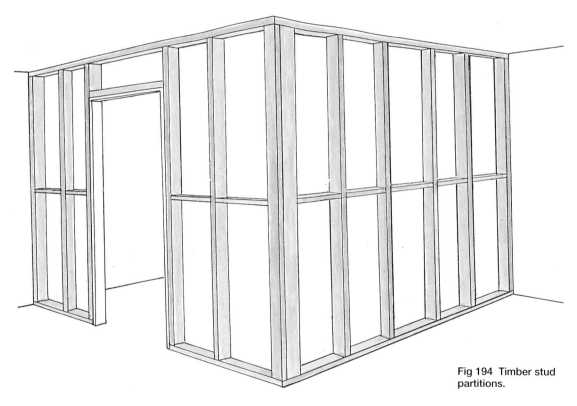

Fig 194 Timber stud partitions.

PHASE 6 – BUILDING (PLASTERING)

It would appear that internal wall and floor finishes have travelled full circle during the past 2,000 years or so, ever since the early Roman and Greek civilizations used gypsum-based mixtures for wall and ceiling coverings, and low-fired red clay terracotta as floor tiles. Today gypsum plays a very prominent role in the building industry and terracotta floor tiles are also incredibly popular with their beautiful warm colour and extraordinary longevity.

Mined in various locations throughout the UK and around the world, gypsum has become very popular throughout the twentieth century to cover the rough surfaces of internal blockwork producing a hard, smooth surface suitable for decorating. Applied in two or three coats, gypsum plaster has superseded sand and lime wall finishes because of its drying out process and the fact that it is less likely to crack. Plasterboards are sheets of gypsum plaster held together by sheets of paper, to prevent cracking, and are almost exclusively used as a covering for ceilings and stud partition walls.

WINDOW BOARDS

When the windows have been installed and the walls built up to the correct levels the window boards can be fitted. For wooden windows the boards will be rebated to fit into the groove at the back of the window and, when ordered, should be

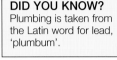

DID YOU KNOW?
Plumbing is taken from the Latin word for lead, 'plumbum'.

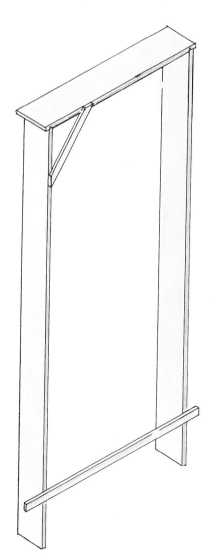

Fig 196 Assemble and square-up softwood door linings before installation.

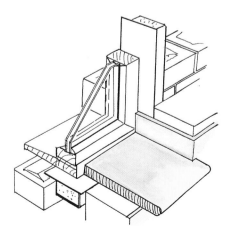

Fig 195 Window boards will be secured to internal walls before plastering.

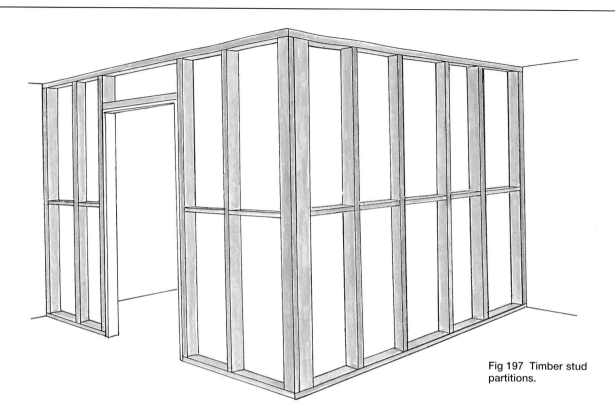

Fig 197 Timber stud partitions.

at least 450mm wider than the window to allow for the window returns. The return is where the window board fits neatly past the blockwork reveals and is plastered around to finish. The boards should be fixed flat and level with the nosed front edge and level with the window line.

STUDWORK WALLS

The design and use of partition walls, particularly at first floor level, has been transformed due to the extensive use of truss rafters in modern house building. At one time all internal walls were load bearing with the first floor walls built in brickwork or blockwork off the ground floor walls, thus carrying the roof load. The job of carrying the roof load has now been taken on by the truss rafters which are designed to span between the opposite external walls of a building, allowing the first floor partitions to simply become room dividers.

Consequently, allowed a free choice of wall position, architects and designers regularly choose lightweight stud partitions and core filled partitions when planning first floor constructions. The lightweight stud partition wall is constructed with 100 × 50mm softwood sawn timbers with a plasterboard skin on both sides. To reduce sound travelling through the partition, when used for bathrooms and toilets, for example, the cavity can be filled with mineral wool slabs or glass fibre quilt.

To build a stud partition the wall positions will be set out and then, following the lines marked on the floor and the ceiling, two horizontal timbers acting as a base plate and a header plate will be fixed. The first, the base plate, is nailed to the flooring along the line of the proposed wall and the second, the header plate, follows the same line, but is fixed to the ceiling. For added strength make sure you nail these timbers through the floorboarding and ceiling boarding into the floor and ceiling joists. The vertical lengths of timber, called studs, can then be erected and secured at 400mm, 450mm or 600mm centres. The spacing of the studs will depend on the thickness of the plasterboard you

STUD PARTITIONS
When you build the stud partitions, try to insert timber fixings from which to hang items; pictures, for example, are usually hung at a certain level, so a timber fixing can be added for this purpose.

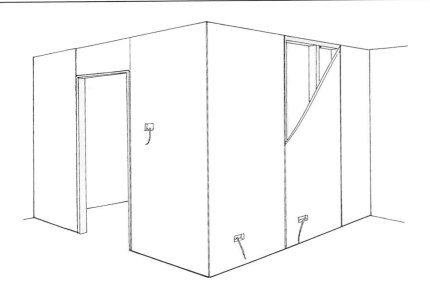

Fig 198 Nail plasterboards to studding, allowing for services such as electrics and plumbing.

use and the width of the boards. For 9.5mm plasterboard the spacing should be no more than 450mm and for 12.55mm plasterboard the spacing should be no more than 600mm. Standard plasterboards are 1200mm wide so the studs should be spaced to allow for the boards to join onto a timber stud.

There will be a third horizontal strip, called a Noggins, which is cut down to fit about mid-height between the vertical studs, which adds rigidity to the framework and provides another fixing point for the plasterboards. Door linings will be built into the structure before the plasterboard is added and electrical cables, boxes and even central heating pipes will all be installed before the walls are covered. All the services to be hidden in stud walls must be secured properly using the appropriate clips and checked to ensure there is no risk of leakage or short circuiting. When the stud walls are completed they can be filled with sound deadening quilt, if required, then covered with plasterboard sheets.

As an added protection, usually to prevent or delay the spread of fire, plasterboards can also be fixed to studwork walls in double layers. The first layer will be nailed to the studs then the second layer, fixed directly over the first layer but with the joints staggered to prevent a weak point, can be added. This type of fire protection may be required when partitioning off part of a garage, for example.

PLASTERING AND SKIMMING

Apart from decorative stone walling and the occasional brick fireplace, we do prefer the majority of our internal walls to be either painted or wallpapered. To provide a suitable surface for these decorating processes the internal walls will be covered with at least two coats of plaster, a base coat to cover the rough surface of the building blocks and a smooth skim coat to finish. Plastering is definitely a highly skilled trade which requires both speed and accuracy.

The first coat, called the floating coat, is applied using a hawk and trowel to a depth of between 10mm and 13mm over the wall. A straight edge will be used to provide a level and even surface and then a special tool called a 'devilled' float will be dragged over the surface of the plaster to provide a key for the finishing coat. The 'devilled' float or trowel has sharp points and provides a gauged effect on the surface of the floating coat of plaster. The base coat will then be allowed to set, but not dry out completely, for about two hours, then the smooth finishing coat can be applied using a float and brush. The brush is used to flick water onto the plaster surface which, when the float is used, produces a hard shiny finish. If the base coat gets too dry it may require

WHAT PLASTER?
Before you order the plastering materials, make sure the plasterer tells you which materials will be best suited to plaster your walls. There are two plastering methods to choose from and both will have a lightweight gypsum plaster finish. The first method will be a gypsum browning base coat and the second will be a sand and cement base coat. The decision about which method is more appropriate will depend upon the surface to be plastered.

dampening before the skim coat can be applied.

To provide a smooth finish on plaster-board walls, a plaster skim coat is applied after the plasterboard joints have been taped. The drying out process for this structure is very short indeed but the drying out of two-coat plaster walls will take a lot longer.

Electrical cables must be in place before the plastering commences, but the fittings will be added after the walls are plastered. Never plaster near or around live electrical wires because the water involved with plastering can have fatal consequences.

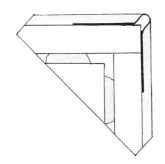

Fig 199 Plasterer's metal corner beading.

Fig 200 Internal walls plastered with a base coat before a final skim coat.

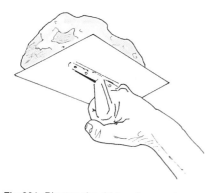

Fig 201 Plaster should be of a good, dropping consistency.

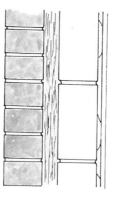

Fig 202 Cavity wall construction with a plaster finish to internal walls.

Fig 203 Stop bead for concrete rendering.

PLASTERING OVER IT?

Plastering walls made from two different materials, perhaps part brickwork and part blockwork, could result in the plaster cracking where the two surfaces join. This is due to the varying movement of the plaster on the different surfaces. To prevent this, a metal lathe can be fixed over the join before the wall is plastered.

ARTEXING

There are usually only two possible choices for the decoration of ceilings. The first is a plaster skim which is then painted and the second, more popular in homes built in the last twenty years, is a white surface coat call Artex. To apply Artex the ceiling boards must be fixed with the ivory face downwards to help the white Artex mixture provide a nice crisp finish. The plasterboard joints are taped and then Artex mixture is applied using a special brush. Then, before it sets, a variety of finishes can be produced, including a stipple-type finish and patterns of all kinds, although your choice is likely to be the same as the finish used on your existing ceilings. It is reasonable to assume that the Artexer will copy your existing style unless he is otherwise directed. If you want to try something different ask to see the different possible patterns available before you go ahead. However, be warned, unlike plastered ceilings where you can easily change the colour, Artexed ceilings are permanent.

One other important point to remember is that the Artexer will expect to Artex the ceilings *after* the electrician has secured the light fittings. The plasterer, on the other hand, where the ceilings are to be skimmed, will apply a plaster skim finish *before* the electrical fittings are in place.

DRY-LINING

Dry-lining is quite popular in domestic construction and is used by some larger housebuilders because its faster drying out period allows for earlier completions. To dry-line walls, as an alternative to plastering, plasterboards are secured to the walls with timber battens or plaster dabs, the board joints are taped and the walls are then ready for decoration, including wallpapering. The boards used for dry-lining have tapered edges where the boards meet and where a joint can be formed using a joint compound, a joint tape and then a slurry finish. This method is an alternative to wet plastering and it does have its benefits. It is clean, the boards can be

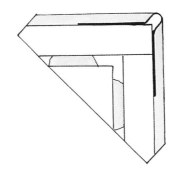

Fig 204 Corner beads for plasterboard partitions.

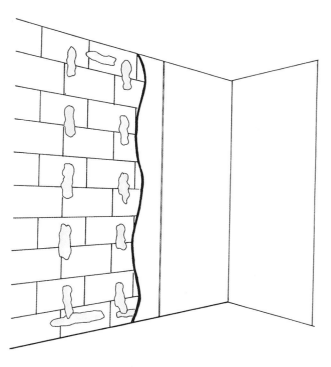

Fig 205 Dry lining plasterboard walls using the dot and dab method.

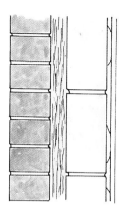

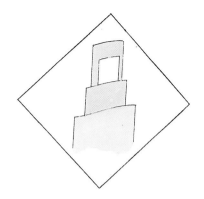

Fig 206 *(far left)* Cavity wall construction with a dry lining finish to internal walls.

Fig 207 *(left)* Plasterboards will be jointed using special compound and tape.

decorated almost immediately and there is no shrinkage resulting in cracks. The process is more expensive than plastering but it does add to the thermal qualities of the walls. If you do plan to dry-line the walls of your extension you will need a specialist installer to carry out the work.

FLOOR SCREEDING

A floor screed is a sand and cement mixture laid to provide a flat and smooth floor finish and is now commonly used in domestic building construction. This type of solid floor, comprising a hardcore base, damp proof membrane and a concrete oversite finished off with a concrete screed, has eliminated the risk of

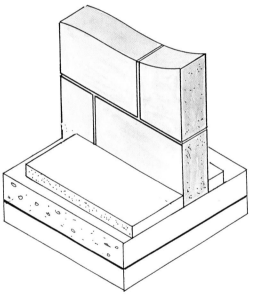

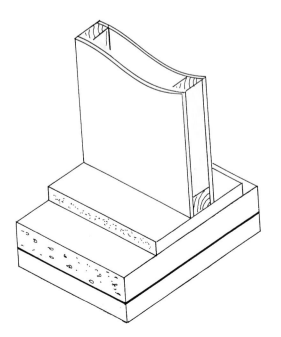

Fig 208 *(above)* Concrete screed around block walls and a concrete sub floor.

Fig 209 *(left)* Timber partitions built off a concrete screed floor.

dry rot and other fungal problems previously attributed to timber floors. A layer of insulation can be installed between the oversite concrete and the floor screed to meet thermal insulation requirements.

A floor screed is a mixture of sharp sand and cement to a ratio of between 1 in 3 and 1 in 4, depending upon the depth of the screed. A screed laid on the oversite concrete should be at least 40mm thick and when it is laid on a layer of insulation, called a floating screed, it should be at least 65mm thick. Where heating cables and service pipes pass through the screed add 10mm of screed to the above thicknesses to avoid cracking.

Laying a smooth, level floor screed on a concrete oversite base is a specialist job, usually carried out by the plasterer after the walls have been plastered. The oversite should be dampened down before a slurry coat of water and cement is applied to provide a bonding mixture; the complete floor area should be laid in a day to avoid joins.

FLOOR TILING

Allow at least a two-week drying out period before laying your floor tiles on a newly laid screed. Before laying your tiles, in fact before laying the floor screed, you must calculate where the finished floor level will be. Then you can calculate the depth of floor screed required. The floor tiles can then be laid, as required, and according to the manufacturer's specifications, but never cut down doors until the finished floor level is clearly established. Skirting boards or skirting tiles will provide

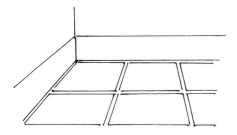

Fig 210 Set floor tiles out from the centre of the room, not from a corner.

the finished edgings and should be added when the floor tiles are down. If your floor includes a hidden manhole cover (inspection chamber lid), then this must also be installed to the finished floor, level with the tiles bedded into the cover. Setting out the tiles should be carried out either from the manhole cover, if there is one, or from a line drawn centrally down the room and then adjusted slightly to reduce the number of cut tiles along the edges.

DRYING OUT

The drying out process of a two-coat plaster wall will depend upon the moisture content of the walls being plastered and the atmospheric moisture. The plasterer will be the best person to advise on how long the walls should be left before decoration. It is likely that a water-based paint, not vinyl, will be best to use at first. Skimmed plasterboards on ceilings and stud partitions, on the other hand, are a much drier product and they will be ready for decoration within a very short period of time.

The urge to occupy your new extension will be delayed while this drying out process is going on, but do not try to dry out the building using heaters if it can possibly be avoided. The natural drying out or curing of water-based surfaces such as floor screeds and plastering provide a more secure and permanent surface if they are allowed to dry at a sensible pace.

TRADESMEN

This phase of work is likely to involve the carpenter, plasterer and Artexer.

- The carpenter: will build studwork walls including door linings. Will add the sills to windows and doors.
- The plasterer: will two-coat plaster walls and skim plasterboards and concrete screed the floor.
- The Artexer: will Artex the ceilings.
- Labour: will tack ceilings and studwork walls with plasterboards.
- Plant hire: a cement mixer for mixing the floor screed. Trestles and planks for the plasterer.

Phase 6 – Building (Plastering)

STEP-BY-STEP SUMMARY

Partitions Boarded

After the stud partitions are built they will need to be covered. Plasterboard is most commonly used, though other decorative sheeting, including timber, can be used.

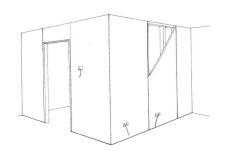

Plastering

The next stage will involve applying the wall covering, plaster or plasterboard. The walls should be clean and made ready with all holes filled in.

Floor Screed

Following on from the walls, the plasterer will lay the floor screed. Make sure damp proofing precautions have been taken and any pipes or cables under the screed are protected against damage.

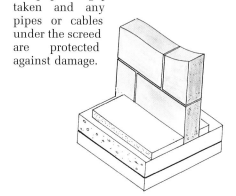

Internal Doors

With the finished floor levels determined, the internal doors can be cut to height. Never do this before you know exactly where the floor levels are and then hang them in their appropriate positions.

Skirting and Architraves

As with the internal doors, the skirtings and architraves can now be fitted.

Sanitary Ware

Any sanitary ware (bath, toilet, basin, bidet etc.) should be secured to walls and floors and connected to the drains ready for the services to be added.

Fig 211 A 'P' trap toilet.

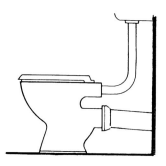

SECOND FIXING

Both the building structure and the aesthetic appearance of the modern domestic dwelling has been brought about because of the demands placed upon designers and architects to conserve energy and provide adequate accommodation for the twentieth century family within an ever reducing space.

The internal finishes and furnishings, however, appear to be going in a totally different direction seeking a style long past. With both Tudor and Elizabethan designs and room layouts admired as 'the way to go', and with antique furniture and trinkets adorning more and more modern homes, the overall picture of progress is becoming confused. The desire to live comfortably has been accompanied by the equally strong desire to display one's wealth and this must be as a result of the housing revolution of the 1970s.

To confuse the picture even more you will find that not all rooms are heading in the same direction. The bathroom styles of the 1960s, 1970s and 1980s are constantly scorned by designers extolling the virtues of pre-war fixtures and fittings, while kitchens are certainly not trying to revert to days gone by. No one, common theme seems to stabilize this overall picture, with each room having almost a complete style of its own. On a positive note this adds colour and splendour to our homes as we head into a new century and can only be considered a definitive move and a testament to past generations whose styles are still popular and much sought after. This emphasis on self-selection permeates a complete range of possibilities, where even the smallest of homes can now reflect a range of choices never before seen in home design.

The style of your new home extension may encompass all these possibilities, or it may simply be an addition to meet increasing needs. Whatever the outcome and the choice of design or decoration, the structural preparations will be the same and must have been followed to provide a base on which you can achieve only the best possible results.

SECOND FIX: ELECTRICS

To complete the contract the electrician will return, when instructed, to add socket and light fittings to the cables fixed earlier in the project. The second fixing will be dependent upon the finishes being planned for walls and ceilings.

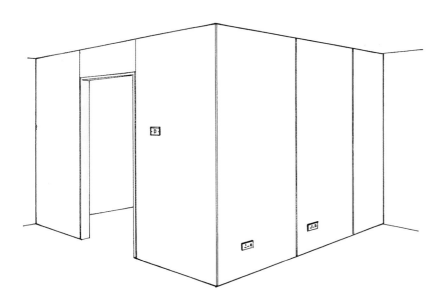

Fig 212 When the boards are secured and the joints taped, the electrical sockets and switches can be fitted.

Second Fixing

Ceiling roses, for light fittings, should be fixed *before* the ceiling is Artexed but *after* the ceiling is plaster skimmed. The process of adding fittings is fairly quick and, once the circuits have been tested, then the job is finished. There is a wide range of fittings and fixtures on the market but it is likely that the majority will complement the existing fittings. Any changes should be discussed with the electrician at the quotation stage, or at least before the second fix is completed. It is not too late to add sockets or lights but there will be a lot of extra repair work to be carried out. Getting it right at the first fix stage is very important.

SECOND FIX: PLUMBING

Second fix plumbing involves quite a lot more work than is required for the electrical fittings, although the process is rather similar. Fixing radiators will be done after the walls have been plastered and, with careful planning, after decorating. Painting behind a radiator is rather tiresome so if you have

decided on materials and colours this is a good time to paint those awkward areas. Fixing washing and toilet facilities will also be carried out after the walls have been plastered. All of these facilities can be tested before the plumber leaves the site, but small leaks do occur, and often within the first couple of days, so contact should be maintained for future reference.

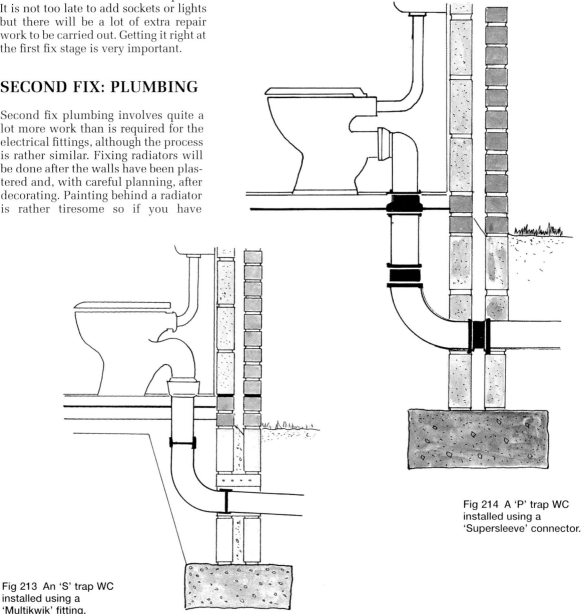

Fig 214 A 'P' trap WC installed using a 'Supersleeve' connector.

Fig 213 An 'S' trap WC installed using a 'Multikwik' fitting.

126

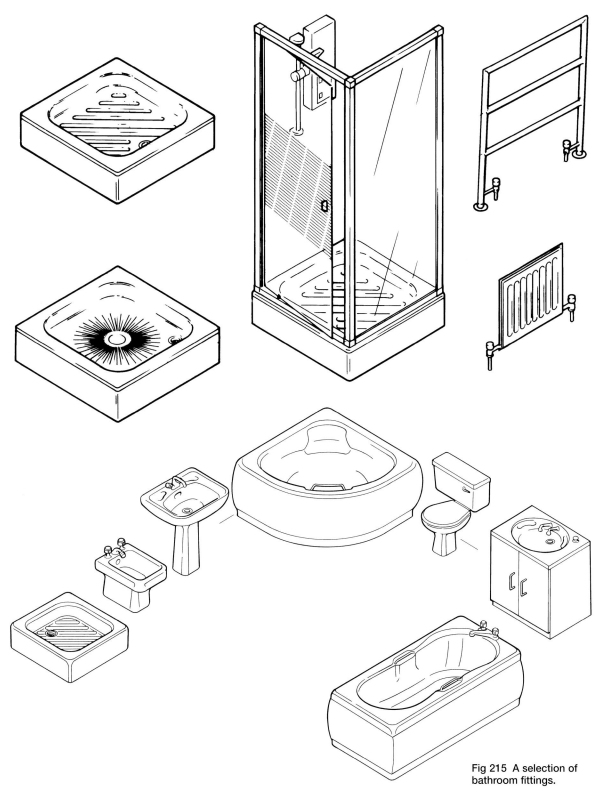

Fig 215 A selection of
bathroom fittings.

Second Fixing

INTERNAL DOORS

The internal door, unlike its external counterpart, is commonly a very lightweight timber door fixed into a softwood door lining, hinged at one side allowing the door to open in the direction best suited for entry or exit. Choices can include single doors, double doors (where the opening is wider) and sliding or folding doors (where access may be limited).

type and strength must meet with requirements.

Finally the ledger and braced door, of a heavy or a lightweight construction, is very much a country-style door and can be installed as a stable door where the top half and bottom half can open independently. These are popular, but not necessarily ideal for privacy or sound proofing. The door selection you make will almost certainly depend upon the existing doors you have,

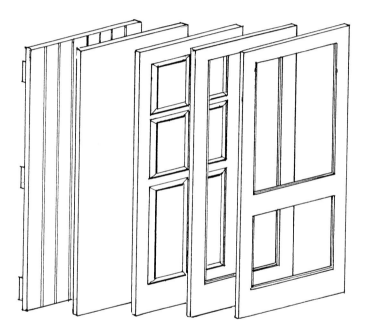

Fig 216 There will be a wide selection of doors to choose from.

The range of internal doors to choose from includes flush doors, panelled doors, glazed doors and ledger and braced doors. With such a wide range to choose from the purchaser is clearly spoiled for choice. Flush doors can have a hardwood veneer finish, removing the requirement for decorating, or have a plain timber finish to be decorated in the colour required. Panelled doors also are available in numerous styles both pre-decorated and plain, or primed, ready for decoration. Glazed doors are heavier and rather more substantial so require more attention when fixing. There are restrictions with glazed doors relevant to the type of glass used and at what height the glass panels start. Accidents can occur with glazed doors so the glass

although matching, especially with veneer doors, may be difficult.

Whatever door style you choose you will have to fix door furniture to the door. Again, as with the doors themselves, it is likely that your choice will reflect your existing door furniture, but the range is extensive. The choice will range from iron through to aluminium and gold – the possibilities are endless and availability often almost instant.

The hanging of the doors is not extremely difficult and should be feasible for an experienced DIYer, otherwise it is a job for the carpenter and is best carried out as almost the final part of the puzzle. Do not fix doors, except perhaps on a new bathroom, until the plasterer and plumber have finished.

SKIRTING BOARDS AND ARCHITRAVES

The decorative timbers used as an edge to doors and walls are called skirtings and architraves and are moulded strips of timber, cut to length and secured using nails or screws. For appearances the skirting boards and architraves are used to hide the rough edges where the plastering reaches the flooring and abuts the door frame or lining.

There are various styles to choose from including Rounded and Splayed, Ogee and Bull-nosed. Skirting boards should be cut tightly into corners and joined with mitred joints for appearances. They can be secured to the walls using cut nails, or drilled and screwed. Around the doors, two vertical lengths of architrave will be cut to go on either side of the door, with a horizontal length above the door intersecting the verticals using a mitred, 45 degree, joint. Accuracy is fairly important, although filling can be added to conceal gaps where architraves are to be painted. A small margin is left around the lining, for appearances.

As a set, architraves and skirtings are usually the same style, although skirting boards fulfil a different function. The architrave is purely for appearances, whereas the skirting board acts as a dividing line between the floor and the walls,

masking the joints between the plastering and the flooring and acting as a buffer to withstand knocks and bumps. Skirting boards and architraves are usually undecorated timbers which are then primed, undercoated and glossed after fixing.

DADO RAILS AND PICTURE RAILS

Common to houses built in the 1930s and 1940s, and now undergoing something of a resurgence, the dado rail is a centrally situated horizontal rail, placed midway between the floor and the ceiling as a design feature, allowing a division between decoration of the upper and lower parts of a wall. The picture rail, on the other hand, is less design orientated and has a more practical use as a rail to hang pictures from and is not, at the moment, part of any new design trend.

HANDRAILS AND BALUSTERS

Where a staircase is to be installed the fittings, handrails, balusters and so on (the stair string may have been fixed earlier), should be added after the plastering has been completed leaving a good surface to adhere to. Adding handrails and other fittings after the wet trades have finished

STAIRCASE FITTINGS
If you are installing a new staircase, make sure any timbers to be visible are protected from damage and splashing. The treads should also be protected during construction using a strip of timber or chipboard.

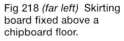

Fig 217 Softwood door linings showing door stop and architrave.

Fig 218 *(far left)* Skirting board fixed above a chipboard floor.

Fig 219 *(left)* Skirting board fixed above a concrete and screed floor.

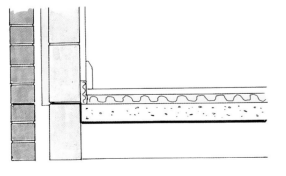
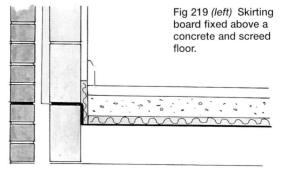

Second Fixing

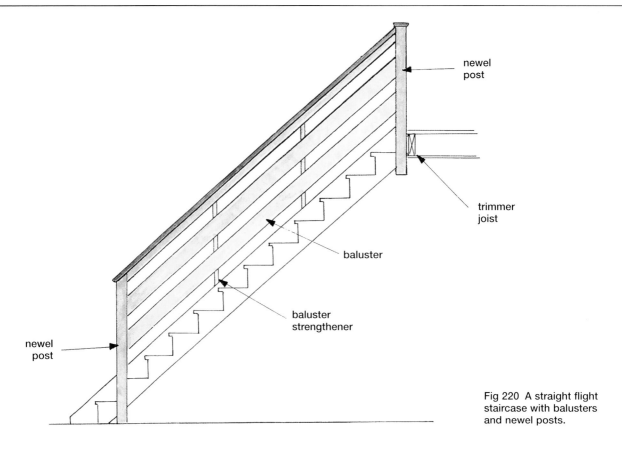

Fig 220 A straight flight staircase with balusters and newel posts.

will reduce the risk of damage to both softwood and hardwood.

The treads should have been protected when the string was installed against damage from traffic and now the handrails and balusters can be fitted. There are several Building Regulation requirements surrounding the height of handrails and the positioning of balusters which must be observed. Handrails will be fitted on one, or both, sides of the staircase, while the open area between the handrail and the stair string is called the balustrade and must be enclosed, either completely or as a barrier to prevent accidents.

Building Regulations require handrails on at least one side for stairs under 1m wide and on both sides for stairs wider than 1m. The position of the handrails should be not less than 840mm, measured from the pitch line of the stairs, and 900mm high measured from the landing. The handrail is a moulded length of timber, softwood or hardwood, secured to the wall or jointed between newel posts. Balusters can be vertically fixed between the handrail and the stairs and are usually square or turned on a lathe. The gap between these balusters should not exceed 90mm. Parallel balusters will run parallel to the handrail and the stair string, with strengtheners in mid-span to provide rigidity, and any gaps that remain should be of a size to prevent children falling or climbing through.

WALL TILING

In bathrooms, kitchens, toilets and utility rooms there may be areas where ceramic or similar wall tiles are to be added to guard against water damage to plastered surfaces. The tiles should be added after the plastering is completed, and fixed according to the manufacturer's specifications.

Fig 221 Wall tiling.

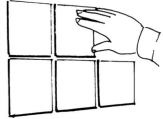

PAINTING AND DECORATING

Newly plastered walls beg to be decorated, but a short drying out period must be allowed before you get the brushes and rollers out. Plastered walls can be decorated with an emulsion-type paint, preferably not a vinyl, within a reasonable time of completion, but hanging wallpaper onto newly plastered walls, especially vinyl-type wallpapers, will prevent the drying out process and should be avoided for several weeks, if not months, after the work is completed. Skirting boards and architraves can be primed, undercoated and glossed before you start on the walls, and the ceilings can be painted. Any stud partition walls constructed from plasterboard and finished with a taped joint of a plaster skim finish, will not require the length of drying out other walls require and can be decorated within days of completion.

TIDYING UP

When the work is finally completed, and all the rubbish can be removed from the site and taken to the local dump, it is likely that there will remain a small quantity of materials, including bricks and tiles. You may not want to keep some of the surplus, but spare bricks and tiles should be set aside in the unfortunate likelihood of accidents occurring and identical replacements being required. How you dispose of the remaining items will depend upon your location, on the possibility of other extension builders

> **DECORATING**
> Before painting or papering newly plastered walls, check with the plasterer about drying out times and ask him to recommend materials to use. There are also a likely to be one or two awkward areas you will want to decorate before fittings, such as radiators and wall lights, are fitted.

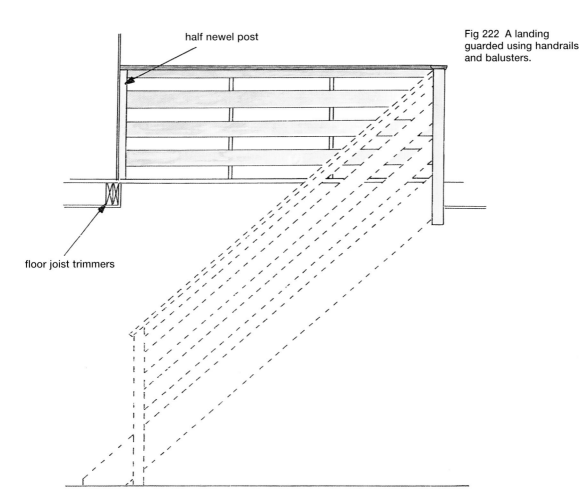

half newel post

floor joist trimmers

Fig 222 A landing guarded using handrails and balusters.

looking to buy surplus items, and on the materials themselves. Spare sand and ballast should not be too difficult to dispose of in the soakaway or somewhere around the garden; perhaps as a base for any paving slabs being added or replaced.

To tidy up the paperwork the Building Control Officer should be notified that the work is now complete and your house insurance should be changed to include the new work. You can also complete the financial calculations that will help you to identify the total expenditure so that you can compare the figures with any quotations you obtained from local builders before the work started.

AND FINALLY ...

I do hope this experience has been an unforgettable and rewarding one and that the savings have been substantial. More importantly, as suggested earlier in this book, I hope you have added the names of several new friends to your Christmas card list.

STEP-BY-STEP SUMMARY

Tiling

Wall tiling around sanitary ware and where electricity boxes are involved, in a kitchen extension for example, can now be added and grouted in.

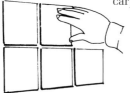

Plumbing

Sanitary ware can now be connected, kitchen fittings can be connected to water pipes and the radiators can also now be

fitted. You may want to decorate behind the radiators before they are fitted.

Electrician

Sockets and light fittings can now be added and connected to the mains.

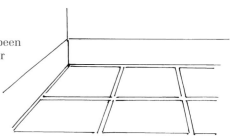

Artexer

The ceilings cannot be Artexed until the ceiling roses are in place. Cover and protect any surfaces from splashes.

Tiling

Floor tiling may have been added when the floor screed was laid or it may be added later. Allow for the floor tile depths when you cut the doors.

Decorator

The final, but equally important, stage: good decorating can hide quite a lot but bad decorating hides nothing. The walls can be decorated and the floors carpeted.

Fig 223 An 'S' trap toilet.

APPENDIX 1

Where a flow chart will help you with an accurate guide to the day-to-day running of your new project, the home extension accounts will help you keep an accurate record of all the expenditures involved. They are useful for preparing estimates of what the project may cost and they are essential if you want to know, accurately, what it did cost to build.

Date :   :   :

Home Extension Accounts

Description
of works   ....

Name   ....

Address   ....
               ....
               ....

PHASE ONE - GROUNDWORKS

£   £

DEMOLITION

Machinery/Plant

Waste disposal

Labour .... hrs/days at .... per hr/day ....

SITE CLEARANCE

Machinery/Plant

Waste disposal

Labour .... hrs/days at .... per hr/day ....

EXCAVATION (inc. drains, soakaways)

Machinery/Plant

Waste disposal

Labour .... hrs/days at .... per hr/day ....

FOUNDATIONS (pre-mixed)

.... cubic metres at £ .... per cubic metre

Machinery/Plant

Labour .... hrs/days at .... per hr/day ....

FOUNDATIONS (on-site mix)

Sand/ballast .... tons at £ .... per ton

Cement .... bags at £ .... per bag

Additives

Machinery/Plant

Labour .... hrs/days at .... per hr/day ....

DRAINAGE (new works)

Materials	Type	Size	Qty	Price	
Pipes					
Connectors					
Fittings					
Fittings					
Gully					
Gully					
Manhole fittings					
Manhole covers					
Pea shingle					
Machinery/Plant					
Labour .... hrs/days at .... per hr/day					

DRAINAGE (old works)

Machinery/Plant	
Labour .... hrs/days at .... per hr/day	

BELOW GROUND MASONRY (inc. manholes)

Build solid walls .... mm thick by .... square metres

Build cavity walls .... mm thick by .... square metres

Materials	Type	Size	Qty	Price	
Bricks					
Bricks					
Blocks					
Blocks					
Lintels					
Sand					
Cement					
Lean mix concrete to fill cavity to ground level					
Machinery/Plant					
Subcontract labour					
Labour .... hrs/days at .... per hr/day					

OVERSITE

Strip oversite area of all vegetative materials.

Well compact .... mm of clean hardcore

Sand blind the hardcore to receive polythene membrane.

Machinery/Plant	
Labour .... hrs/days at .... per hr/day	

DAMP PROOF MEMBRANE

Supply and fit .... gauge polythene membrane

Supply and apply liquid membrane to manufacturer's specifications

Machinery/Plant

Labour .... hrs/days at .... per hr/day ....

CONCRETE SLAB (pre-mixed) inc. paths, etc.

Lay .... mm thick concrete slab .... square metres. Mix.

Cubic metres required .... at £ .... per cubic metre

Machinery/Plant

Labour .... hrs/days at .... per hr/day ....

CONCRETE SLAB (on-site mix) inc. paths, etc.

Lay .... mm thick concrete slab .... square metres

Sand/ballast .... tons at £ .... per ton

Cement .... bags at £ .... per bag

Additives

Machinery/Plant

Labour .... hrs/days at .... per hr/day ....

OTHER GROUNDWORKS

1)

Materials

Machinery/Plant

Labour .... hrs/days at .... per hr/day ....

2)

Materials

Machinery/Plant

Labour .... hrs/days at .... per hr/day ....

PHASE TWO - STRUCTURE

WALLS

Cavity walls .... m² outer leaf .... m² inner leaf

Solid/single skin walls .... mm thick by .... square metres

Materials	Type	Size	Qty	Price	
Bricks					
Bricks					

Continued on next page

	Size	Qty	Price	£	£
Blocks					
Blocks					
Stone					
Stone					
Cavity fill					
Cavity fill					
Wall ties					
DPC					
Air vents					
Flue linings					
Flue bends					
Pots, chimney					
Fireback					
Sand					
Sand					
Cement					
Additives					

WINDOWS

Relocate .... no. windows into new/existing works

Relocate .... no. external door frames into new/existing works

Relocate .... no. garage door frames into new/existing works

New window frames Type/Cat No.	Size	Qty	Price	
1)				
2)				
3)				
4)				
5)				
6)				
New external door/patio frames Type/Cat no.				
1)				
2)				
3)				
New garage door frames Type/Cat no.				
Single				
Double				
Frame ties				
DPC				
Lintels				
Lintels				

Continued on next page

	Size	Qty	Price	£	£
Lintels					
Lintels					
Lintels					
Lintels					
Subcontractors (bricklayers, etc)					
Machinery/Plant (scaffold, etc.)					
Labour .... hrs/days at .... per hr/day					

WALL PLATES (a)

Secure new .... mm by .... mm wall plate to new/existing works

Fix .... no. bolts size .... at .... mm centres to receive joists

Bed .... mm by .... mm wall plate onto new/existing walls and

secure with mild steel straps at .... mm centres

Materials (wall plate, bolts, etc.)

JOIST HANGERS (b)

Securely fix .... no. joist hangers type .... to wall plates or build into

new/existing walls to receive joists at .... mm centres

Materials (inc. fixings)

CAVITY TRAYS (c)

Instal type .... cavity tray to manufacturer's specifications

to discharge above flashings where roof abutments occur.

Materials

Machinery/Plant (a, b & c)

Labour (a, b & c) .... hrs/days at .... per hr/day ....

OTHER STRUCTURAL WORKS (inc. manholes, piers, etc.)

1)

2)

Materials

Machinery/Plant

Labour .... hrs/days at .... per hr/day

				£	£

METALWORKS

				£	£
Instal .... kg beam/RSJ on padstones/plates as specified					
Length .... Size					
Other metalworks (inc. posts, etc.)					
Materials					
Labour .... hrs/days at .... per hr/day					

PHASE THREE
ROOFING, FLOORING & CARPENTRY

FLAT ROOF

Materials	Size	Length	Qty	Price	
Decking					
Firring Pieces					
Joists					
Tilt fillet					
Fascia board					
Soffit board					
Herringbone					
Noggins					
Fixings (inc. vents)					
Machinery/Plant					
Labour .... hrs/days at .... per hr/day					

SLOPING/PITCHED ROOF

Materials	Size	Length	Qty	Price	
Ceiling Joists					
Rafters					
Ridge board					
Hip board					
Valley boards					
Barge boards					
Fascia board					
Soffit board					
Purlins					
Binders					
Struts					
Bracing					
Noggins					
Fixings (inc. vents)					
Subcontract labour					
Machinery/Plant					
Labour .... hrs/days at .... per hr/day					

TIMBER FLOORS

Materials	Size	Length	Qty	Price	£
Joists					
Decking					
Herringbone					
Noggins					
Fixings					
Subcontract labour					
Machinery/Plant					
Labour .... hrs/days at .... per hr/day					

STAIRCASE

Straight   Qtr Landing   Half Landing   Kite Wind   Spiral

Goings .... at .... mm   Risers .... at .... mm

Materials	Type/Style	Size	Qty	Price	£
Flight					
Handrail					
Balusters					
Newels					
Newel caps					
Landing					
Fixings					
Subcontract manufacturer					
Subcontract labour					
Machinery/Plant					
Labour .... hrs/days at .... per hr/day					

STUDWORKS (inc. svp boxing/hatches, etc.)

Materials	Size	Length	Qty	Price	£
Timber					
Insulation					
Fixings					
Machinery/Plant					
Labour .... hrs/days at .... per hr/day					

INSULATION

Provide .... mm thick insulation to floors, ceilings and voids where specified

Materials	£
Labour .... hrs/days at .... per hr/day	

VALLEYS

Provide and fix Code .... lead .... mm wide or pre-formed valleys
prior to roof coverings

Materials

Labour (see also Plumber) .... hrs/days at .... per hr/day ....

GUTTERING

Provide and fix .... mm guttering to discharge into new/existing
soakaways/drains

Materials	Size	Length	Qty	Price
Channel				
Downpipe				
Ch. brackets				
D/p brackets				
Joints				
Stop ends etc.				
Ch. bends				
D/p bends				
Fixings				

Machinery/Plant

Labour .... hrs/days at .... per hr/day ....

FELT ROOFING

Lay on existing decking 2/3 layers of built up bitumen felt roofing

with reflective chippings bedded in bitumen to finish

All in accordance with CP114 Pt 3 1970,

and materials in accordance with BS 747 (1977)

Roofing .... m² Welted drip .... m Upstands .... m

Dressed into walls below cavity trays .... m

Materials

Machinery/Plant

Labour .... hrs/days at .... per hr/day ....

Subcontractor (supply and fix/labour only)

ROOFING   TILE/SLATE

Materials	Type/Style	Size	Qty	Price
Tiles/slates				
Tiles/slates				
Eaves tiles				
1½ tiles				
Hip tiles				
Valley tiles				
Ridge tiles				
Bonnet hips				

Continued on next page

	Size	Qty	Price	£	£
Ridge vents					
svp flashing					
Battens					
Roof felt					
Verge boards					
Sand/cement					
Machinery/Plant					
Labour .... hrs/days at .... per hr/day					
Subcontractor (supply and fix/labour only)					

ROOF TRUSSES

Supply and fit .... no. manufactured roof trusses

at .... mm centres. Supplied by

Materials (inc. fixings)

Machinery/Plant

Labour .... hrs/days at .... per hr/day ....

ROOFLIGHTS (Velux, etc.)

Supply and fit .... no. .... rooflights and flashings

Materials

Machinery/Plant

Labour .... hrs/days at .... per hr/day ....

PHASE FOUR
FIRST FIX & FINISHING TRADES

CARPENTRY (pre plasterer)

Materials	Type	Size	Length	Qty	Price	
Window boards						
Door linings						
Door linings						
Noggins						
Hatch lining						
Hatchway						
Fixings for appliances						
Fixings						
Labour .... hrs/days at .... per hr/day						

PLASTERER

Materials	Size	Qty	Price	
Plasterboards white/grey				
ditto				
Foil-backed pl.bd. white/grey				
Insulation board white/grey				
Carlite bonding				
Carlite browning				
Carlite finish				
Board finish				
Syrapite				
Soft sand				
Sharp sand				
Portland cement				
Angle beads				
Stop beads				
Expanded metal strip				
Expanded metal sheet				
Artex				
Coving				
Arches				
Pebble dash				
Decorative finishes				
Fixings				

Machinery/Plant

Labour .... hrs/days at .... per hr/day ....

PLASTERER (subcontract)

Tacking. .... m² ceiling tacking

Tacking. .... m² studwork tacking

Walls. Sand/cement & finish .... m² at £................. per m²

Walls. Bonding/Browning & finish .... m² at £ .... per m²

Board skimming. .... m² at £ .... per m²

Artex. .... m² at £ .... per m²

Walls. Sand/cement render. .... m² at £ .... per m²

Pebbledash rendering. .... m² at £ .... per m²

Patching. .... m² at £ .... per m²

Window/door reveals. .... m run at .... per m run

Floor screed. Sand/cement. .... m² at £ .... per m²

Other

Machinery/Plant

WALL & FLOOR TILING

Supply and fix .... m² wall tiles inc. adhesive and grout

Supply and fix .... m² floor tiles inc. adhesive (grout if required)

Subcontract (supply & fix/labour only)

Labour .... hrs/days at .... per hr/day ....

PLUMBING

Hot/cold water systems and drainage (inc. leadworks)

Materials	Size	Length	Qty	Price	

Central heating system

Materials	Size	Length	Qty	Price	

Machinery/Plant

Subcontract (supply & fix/labour only)

Labour .... hrs/days at .... per hr/day ....

SANITARY WARE (inc. sinks)

ELECTRICAL WORKS

Materials	Size	Length	Qty	Price	

Machinery/Plant

Subcontract (supply & fix/labour only)

Labour .... hrs/days at .... per hr/day ....

GLAZING

	£	£
Supply and fit to openings .... mm clear glass by .... m²		
Supply and fit to openings .... mm obscure glass by .... m²		
Supply and fit to openings .... mm safety glass by .... m²		
Other		
Subcontractor (supply & fit/labour only)		
Labour .... hrs/days at .... per hr/day		

PHASE FIVE - FINISHINGS

CARPENTRY (doors)

Supply and fit into frames. .... no. External doors. .... no. Patio doors.
.... no. Garage doors. .... no. Fire doors.

Materials	Type/Cat no.	Size	Qty	Price	
Self closers for fire doors					
Night vents to patio doors					
Crickle vents					
Furniture					
Butts					
Fixings/fittings					
Labour .... hrs/days at .... per hr/day					

CARPENTRY (2nd fix)

Materials	Size	Length	Qty	Price	
Skirting board					
Architrave					
Door stop					
Beading					
Loft hatch					
Hatch doors					
Fixings					

Continued on next page

Machinery/Plant	
Labour .... hrs/days at .... per hr/day	

CARPENTRY (jobbing)

Tank stands, hatchways, templates, etc.

Materials	Size	Length	Qty	Price	

Labour .... hrs/days at .... per hr/day	

DECORATION

Machinery/Plant	
Labour .... hrs/days at .... per hr/day	
Subcontract (supply & fix/labour only)	

JOBBING (inc. access, strip roof, etc.)

1)	
2)	
3)	
4)	
Machinery/Plant	

Labour	1) .... hrs/days at .... per hr/day	
	2) .... hrs/days at .... per hr/day	
	3) .... hrs/days at .... per hr/day	
	4) .... hrs/days at .... per hr/day	

SUMMARY

Item	Materials	Mach/Plant	Waste Disp.	Labour	Sub Con. Labour	Sub Con. Sup. & Fix
Demolition						
Site Clearance						
Excavation						
Foundations (premix)						
Foundations (on-site mix)						
Drainage (new works)						
Drainage (old works)						
Below ground masonry						
Oversite						
Damp proof membrane						
Concrete slab (pre mix)						
Concrete slab (on-site mix)						
Other groundworks						
Walls						
Windows						
Wall plates						
Joist hangers						
Cavity trays						
Other structural work						
Metalworks						
Flat roof						
Sloping/Pitched roof						
Timber floors						
Staircase						
Studworks						
Insulation						
Valleys						
Guttering						
Felt roofing						
Roofing. Tile/Slate						
Roof trusses						
Rooflights						
Carpentry (pre plaster)						
Plasterer						
Plasterer (subcontract)						
Wall & floor tiling						
Plumbing						
Sanitary ware						
Electrical works						
Glazing						
Carpentry (doors)						
Carpentry (2nd fix)						
Carpentry (jobbing)						
Sub Totals						

Continued on next page

Sub Totals						

Item	Materials	Mach/Plant	Waste Disp.	Labour	Sub Con. Labour	Sub Con. Sup & Fix
Decorating						
Jobbing 1						
Jobbing 2						
Jobbing 3						
Jobbing 4						

Sub Totals

Sub Totals £

Materials	
Machinery/Plant	
Waste disposal	
Labour	
Subcontract labour	
Subcontract supply/fix	

Total

Notes

APPENDIX 2

PREPARING A FLOW CHART

In the building industry, good planning is essential and a flow chart is a record of what has happened, and, when used correctly, will give an accurate forecast of what is about to happen.

Preparing a flow chart simply means listing the dates that you placed orders for plant or materials and the dates that they were delivered to site, then matching them with the starting dates agreed with the tradesmen. Planning these two factors correctly is as important, if not more so, with a smaller project as it is with a larger one. The turnover of material is greater and the time spent on site by the tradesmen is shorter, thus allowing only a small window for error. As the project progresses you will find that more and more of the important facts on your flow chart become history, but that only means they can no longer be changed.

The only important fact not included in the flow chart is finance. For good reason it may not be relevant to ordering materials, arranging delivery dates or agreeing starting dates with tradesmen, but it will play a very important role in the continued running and successful outcome of your project. From the date you submitted your building plans to the Local Authority for approval until the time the project is completed these charts will be a day-to-day diary and remain as a reference or guide for any projects you may plan in the future.

Fig 224 A flow chart is a day-to-day diary recording orders, deliveries and starting dates for tradesmen.

Flow Chart - Check List

	Planning Permission	Applied for : 17th June '99	Approved : 2nd Oct '99	Fees : £ 95.00p		
Building Plans					Architect fees : £ 650	
	Building Regulations	Applied for : 9th Aug '99	Approved : 29th Sept '99	Fees : £ 88.13p	Total Cost	£ 1097.50p
					Inspection fee : £ 264.37	

Groundworks

		Excavation works	Demolition works	Drainage trenches		Cost
	Materials & Plant			Labour/Tradesmen		
Item	Date ordered	Date delivered	Name	Started	Finished	1) £ 130.00p
1) Tipper Lorry	18th Oct '99	19th Oct '99	4) Bill	19th Oct '99	26th Oct '99	2) £ 12.50p
2) Angle Grinder	18th Oct '99	19th Oct '99	5) Andrew	19th Oct '99	26th Oct '99	3) £ 75.00p
3) Tipping Fees	25th Oct '99	25th Oct '99				4) £ 200.00p
						5) £ 100.00p
					Total	£ 517.50p

Below Ground Masonry

		Concrete foundations	Walls to dpc level	Oversite area		Cost
	Materials & Plant			Labour/Tradesmen		
Item	Date Ordered	Date delivered	Name	Started	Finished	
					Total	

Flow Chart – Check List

Building Plans

Planning Permission	Applied for :	Approved :	Fees :
Building Regulations	Applied for :	Approved	Fees :

Architect fees :   Total cost :

Groundworks

Materials & Plant		Excavation works	Demolition works	Drainage trenches	
Item	Date ordered	Date delivered	Name		Cost

Labour/Tradesmen			
Started	Finished		Cost

Total

Below Ground Masonry

Materials & Plant		Concrete foundations	Walls to dpc level	Oversite area	
Item	Date Ordered	Date delivered	Name		Cost

Labour/Tradesmen			
Started	Finished		Cost

Total

Walls

Materials & Plant		Cavity walls	Partition walls	Windows & Doors	Labour/Tradesmen			Cost
Item	Date ordered	Date delivered	Name		Started	Finished		
							Total	

Roofing Works

Materials & Plant		Flat roof	Pitched roof	Cavity trays/Leadwork	Labour/Tradesmen			Cost
Item	Date ordered	Date delivered	Name		Started	Finished		
							Total	

Appendix 2: Flow Chart – Check List

	Roof coverings						
Materials & Plant	Felt roof	Tiled roof	Gutterings			Labour/Tradesmen	Cost
Item	Date ordered	Date delivered	Name	Started	Finished		
						Total	

	First Fix Carpentry						
Materials & Plant	Timber floors	Partition walls	Stairs			Labour/Tradesmen	Cost
Item	Date ordered	Date delivered	Name	Started	Finished		
						Total	

	Services						
Materials & Plant	Electrics	Plumbing	Glazing			Labour/Tradesmen	Cost
Item	Date ordered	Date delivered	Name	Started	Finished		
						Total	

Plastering

Item	Materials & Plant		Walls & ceilings	Screeding	Artexing	Labour/Tradesmen			Cost
	Date ordered		Date delivered	Name		Started	Finished		
							Total		

Second Fix Carpentry

Item	Materials & Plant		Doors	Architraves & Skirting	Stairs	Labour/Tradesmen			Cost
	Date ordered		Date delivered	Name		Started	Finished		
							Total		

Finishings

Item	Materials & Plant		Second fixing	Decorations	Snagging	Labour/Tradesmen			Cost
	Date ordered		Date delivered	Name		Started	Finished		
							Total		

153

USEFUL ADDRESSES

This short list of manufacturers and stockists has been assembled to help and guide you in any product selection you make when planning and building your new home extension. If you have any questions about either installation or availability of their products, I am sure they will be happy to provide you with all the information you require.

Concrete Blocks

Tarmac Topblock Ltd, Wolverhampton. (01902) 754131

RMC Concrete Products Ltd, Buxton, Derbyshire. (01298) 22244

Insulation Blocks

Thermalite Blocks, Marley Building Products, Sevenoaks. (0990) 620900

H & H Celcon Blocks, Sevenoaks. (01732) 880580

Drainage (Clay)

Naylor Clayware Ltd, Barnsley. (01226) 790591

Hepworth Building Products, Sheffield. (01226) 763561

Drainage (Plastic)

Polypipe plc, Doncaster. (01709) 770000

Caradon Terrain Ltd, Aylesford, Kent. (01622) 717811

Lintels

Dorman Long Ltd, British Steel Lintels, Newport, Gwent. (01633) 244000

Caradon Catnic Ltd, Caerphilly, Wales. (01222) 885955

Expamet Building Products, Hartlepool. (01429) 866688

Bitumen Damp Proof Membrane

Feb Ltd, Swinton, Manchester. (0161) 7947411

Insulating DPC

TDI (UK) Ltd, Matlock, Derbyshire. (01629) 733177

Thermabate, RMC Products Ltd, Wakefield. (01924) 362081

Roof Trusses

Gang Nail Systems Ltd, Aldershot, Hants. (01252) 334562

Jewsons Builders Merchants, your local branch.

Wykes Building Supplies, your local branch.

Roofing Felt

Monarflex, St Albans, Herts. (01727) 830116

Roof Tiles

Marley Building Products, Sevenoaks. (0990) 626900

Plasterboard

Lafarge Plasterboard, Bristol, Avon. (01275) 375281

Cavity Insulation

Rockwool Ltd, Pencoed, Bridgend. (01656) 862621

Kingspan, Leominster, Herefordshire. (01544) 388601

Insulation Quilt

Superglass Insulation Ltd, Stirling, Scotland. (01786) 451170

Floor Insulation

Kay-Metzler Ltd, Wickford, Essex. (01268) 766301

Plumbing

Polypipe plc, Doncaster. (01709) 770000

Radiators

Caradon Stelrad Ltd, Swinton, Mexborough. (01709) 578878

Terracotta Floor Tiles

Fired Earth plc, Adderbury, Oxon. (01295) 812088

Septic Tanks and Sewage Treatment

Klargester Environmental Engineering Ltd, Bucks. (01296) 633000

Entec (Pollution Control) Ltd, Andover, Hants. (01264) 357666

Rainwater Goods

Hunter Plastics Ltd, London SE28. (0181) 855 9851

Hepworth Building Products, Sheffield. (01226) 763561

Doors and Windows

Dale Joinery, Rochdale, Lancs. (01706) 667700

Staircases

Meer End Woodturners, Kenilworth, Warks. (01676) 534226

FURTHER READING

Planning: A Guide to Householders
Dept of Environment
Publications Despatch Centre
Blackhorse Road
London SE99 6TT
Tel: (0181) 694 0099

Building Information Booklets
Building Research Establishment
HMSO Publications Centre
Tel: (0171) 242 6393

National House Building Council
Amersham, Bucks.
Tel: (01494) 434477

ADDRESS BOOK OF TRADESMEN

Full details of tradesmen used. (For future reference.)

Architect/Draughtsman

Name . . . .

Address . . . .

. . . .

Tel. no. . . . .

Bricklayer

Name . . . .

Address . . . .

. . . .

Tel. no. . . . .

Carpenter

Name . . . .

Address . . . .

. . . .

Tel. no. . . . .

Felt Roofer

Name . . . .

Address . . . .

. . . .

Tel. no. . . . .

Roof Tiler

Name . . . .

Address . . . .

. . . .

Tel. no. . . . .

Electrician

Name . . . .

Address . . . .

. . . .

Tel. no. . . . .

Plumber

Name . . . .

Address . . . .

. . . .

Tel. no. . . . .

Glazier

Name . . . .

Address . . . .

. . . .

Tel. no. . . . .

Scaffolder

Name . . . .

Address . . . .

. . . .

Tel. no. . . . .

Plasterer

Name . . . .

Address . . . .

. . . .

Tel. no. . . . .

Artexer

Name . . . .

Address . . . .

. . . .

Tel. no. . . . .

Painter/Decorator

Name . . . .

Address . . . .

. . . .

Tel. no. . . . .

INDEX

air bricks 55–6, 60, 100
air changes 45
aluminium 51
angle grinder 67, 76, 85, 97
Approved Inspector 18, 21
architect 22–3, 25, 118, 125
architraves 129
area plan 17–18, 20, 22
Areas of Outstanding Natural
    Beauty 14
Artex 120–2, 132
asphalt 81–2

back inlet gully 45–6
background ventilation 70
ballast 51, 53
barge boards 90
bath 45
bathroom suites 99, 127
below ground level 51
binders 87–8
bitumen felt 78, 81
block plan 17–18, 20, 22
blockage 45
blocks 28–9, 32, 34, 39, 50–51,
    53, 56, 63, 69
boundary 51
bricklayer 50, 53, 64, 70, 72, 75,
    76–8, 84, 97, 106
bricks 25, 28–9, 32, 34, 39–40,
    46, 50–1, 53, 62, 69, 100
British Standard CP 144 Part 4 82
British Standards 41
builder's line 50, 54
builders' merchants 25
builder's square 35–6
Building Inspector 40, 42, 44, 47,
    49, 51, 68
Building Notice 18, 21
Building Regulations 13, 18, 21,
    22–3, 35, 41, 42–4, 47, 56, 63,
    65, 68, 70, 78, 80, 95, 99, 102,
    104, 130

carpenter 72, 76, 78, 84, 96, 98,
    106, 123
cavity tray 68, 81–2
cavity wall construction 62–3, 78
cavity wall insulation 63–4, 76
ceilings 108, 117, 126
    boards 78
    joists 86–7
    plastered 111

tacking 111
cement 28, 63–4, 122–3
cement mixer 30–1, 51, 76, 106,
    126
central heating pipes 119
cesspool 43, 47
channel junctions 47
channels 46
chipboard 81, 84, 100–1
clay ducting 55
colorant 64
concrete 35, 37–8, 43–4, 46, 49
    blocks 51, 63
    floor insulation 56–7
    foundations 50–3, 58–9
    lintels 54
    oversite 55, 58, 62
    rendering 63
    screed 57
    tiles 87
condensation 65, 80, 86
Conservation Area 13
copper nails 95
CORGI registered 110
crane-operated 32
cross ventilation 80, 94, 95

dado rail 129
damp proof:
    course 36–7, 54–7, 65–7, 101
    level 53, 62
    membrane 52, 56–8, 101, 123
datum peg 35–6
day rate 26, 97
decking 78, 80–2, 99–100
deliveries 26–7, 32
deposit 27
do-it-yourself enthusiast 30
door frames 70, 72, 76
door linings 104, 106, 117, 119,
    128
doors 63, 65, 68, 70, 82, 104,
    124
double glazing 25, 99, 114
downpipes 103
drain test 44
drainage 41, 43, 62
    pipes 32, 38, 52
    trenches 38
    works 78
drains 41, 43, 99
draughtsman 22–3
drills 31

dropping zone 32
dry lining 121
dry rot 54, 123
drying out 123

eaves 80–1
eaves tiles 94
eaves ventilators 80
electrical cables 119–20
electrician 25, 99, 121, 125, 131
electricity:
    cable 38
    circuit breakers 110
    earth 109
    first fix 109
    light fittings 109
    sockets 110
elevation plan, existing 17–19, 21
    proposed 17–19, 22
energy:
    conservation 86
    saving 63–4
engineering bricks 69
English bond 46
excavations 32–4, 37, 40, 50, 78
existing elevation plan 17–19, 21
expanded polystyrene 64
extensions, single-storey 14, 78,
    84, 97
    two-storey 14, 16, 85, 97
extras 27

fascia board 91, 96–7
felt roofer 81, 84–5
felt, roofing 81, 85
fire:
    precautions 70
    protection 119
    resistance 112
firring section 80
flat roof:
    construction 78
    insulation 102
floor:
    boards 54, 100–1
    floating 101
    insulation 57, 101
    joists 83, 100
    screeding 122
    tiles 123
    tiling 123
    wooden 54
flow chart 24–5, 31, 38, 99

fly tipping 35
footings 38, 49–50, 53
foul waste 41–2
foundations 35, 37, 40, 43–4, 76, 104
   strip 51–2
   trench fill 51–2
   trenches 34–6
frame ties 71–2, 104
   galvanized 71
   stainless steel 71
frost damage 58

gable end 87
gable end rafters 76
gable ladder 90, 98
galvanized restraint straps 75–6
garden shed 28
gauge rod 50, 53, 60
gradient board 44
green belt 11
ground level 44
groundworks 26
gutter brackets 96
guttering 41, 90–1, 94, 96, 98
gypsum plaster 117

habitable room 45
hand basin 45
handmade clay tiles 86
hangers 87
hard hats 31
hardcore 52, 56–7, 60–1
hardwood sill 72
health and safety 41
heat loss 63, 66
herringbone struts 82, 85
hip boards 89
hip tiles 89, 93
hipped roof 78, 87, 89, 96

imposed load 69
indents 67, 77
inspection chambers
   (manholes) 38, 44–6
   plastic 46
inspection fee 21
Institute of Electrical Engineers 109
insulation 64, 78, 80, 100, 112–13, 123
   batts 64
   blocks 63
   loose fill 112
   vermiculite 113
insurance 31
internal door 128
internal load-bearing walls 104

internal non-load-bearing walls 104
iron oxides 51

jack rafters 89
JCBs 30, 38
joists 86
   general stress grade 78
   hangers 80
   machine general stress grade 78
   special stress grade 78

kitchen sink 45

land drains 47
lavatory basin 45
layer boards 89
layout plan 17–19, 22
lead 51
lead flashing 83, 97
leakage 44
lime 51, 63
   non-hydraulic 63
   semi-hydraulic 63
limestone 81
lintels 47–8, 53, 65, 68, 82
liquid bitumen 57
Listed building 13
Local Development Plan 11
loft hatch 111

masterboard 91
mechanical electric ventilation 45
metreage rate 26
mild steel restraint straps 80, 83, 86, 96
mineral fibre 57
mortar 51, 63, 67–8, 71, 83

National Inspection Council for
   Electrical Installation 110
National Parks 14
natural stone 63
natural ventilation 59
neighbours 28
noggins 119
noxious gases 45, 48

Ordnance Survey 22
overshadowing 16
oversite 34, 56, 123

'P' trap 45, 48
padstones 69
pantiles 93
partition walls 100

pea shingle 38, 43
permeability 38
Permitted Development 11, 14
pickaxe 38
picture rail 129
pitched roof construction 86
plan fee 18, 21
planks 124
Planning Department 11, 17, 62, 70
Planning Permission 13–4, 22–4
Planning Register 17
plant hire 30, 76, 84–5, 106, 124
plaster 28, 120
plasterboards 117, 119, 131
plastered 101, 126
plasterer 120–1, 128
plastering 119–20
plasticizer 64
plumber 99, 126, 128, 131
plumber first fix 110
plumbing 126
   cold water supply 110
   first fix 110
   flush toilet 108
   gas supply 110
   hot water supply 110
   radiators 110, 126, 131
plywood 81, 101,
Portland cement 51, 63
pre-mixed concrete 32, 51, 53
preservative 92, 96
pricework 26, 97
profile boards 35–7, 39, 50
profiles 67
protective clothing 31
protective sheets 29, 32
public liability insurance 31, 40
public sewer 43
purlins 87, 97

quotations 25–7, 31, 38, 86, 97, 126

'R' values 65, 68
raft foundation 52
rafters 76, 86–7, 91, 97–8
reflective stone chippings 82, 85
reinforced mesh 52
reinforced slab 52
reveals 65, 118
ridge board 87, 89
rising damp 101
rodding eye 45
rolled steel joists (RSJs) 68, 106
roof:
   battens 92
   close-coupled 87

Index

cold 78, 82, 102
  cut 86
  felts 82
  lean-to 87
  mono-pitched 86
  thatched 86
  tiler 97
  tiles 25, 56, 63
  void 80
  warm 78, 82, 102
rubbish 32, 34, 40, 131

'S' trap 45, 48
safety 31–2, 39
sand 28–9, 33, 37, 39, 56, 63–4,
  122
sarking felt 92
scaffold 29–31, 72–8, 85, 98–9
  hire 72, 76
  independent 72
  pole 72
  putlog 72
Schedule 1 18
screed 101, 122
Secretary of State 17
section plan 18, 20, 22
septic tank 43, 47
sewage treatment plants 43
sewers 41, 43
sharp sand 56
shovels 31
silicates 51
site 32
skimming 119–20
skip hire 34
skirting board 101, 123–4, 129
slates 94
sleeper wall 56, 99–100
soakaway 38–9, 47–8, 99, 106,
  131
soffit 80, 90–1, 96–7
soffit vents 91
soil vent pipe 46, 92
spirit level 50
stage payments 27

staircase 99, 102–3, 106
stairs:
  balusters 104, 129–30
  goings 104
  handrail 104, 106, 129–30
  headroom 102, 103
  newel posts 104, 106, 130
  pitch 102, 104
  riser 104
  string 102
  tread 102
steel plates 69
storage 28
straight edge 50
string line 44
structural calculations 21, 69, 89
stud partitions 117–18, 124
studwork walls 118
sub soil 34
supply-and-fix 25
surface water 38, 41–3
suspended floors 54–5, 83, 100–1

tanking 57
tarpaulins 30–2, 34
templates 71–2, 76
theft 28
thermal bridging 64–6
thermal conductivity 64
tile fasçade 63
tiles 93
tilt fillet 81
timber floor joists 100
timber floors 100
timber joists 78
tipper lorries 30, 35, 38
tipping charges 35
top soil 32–4, 39, 56
tradesmen 25–7, 29, 36, 60, 76,
  78, 84, 96, 99, 106, 123
trenches 38–40, 47
trestles 30–1, 124
trickle vents 59, 68
Trinidad Asphalt Lake 81
triple glazing 70

truss rafters 86, 88

'U' values 63
ultraviolet radiation 82
underfloor ventilation 55
Universal Beam 68
upvc pipes 43
urban sprawl 11

valleys 89–90, 92, 94
  boards 89
  liners 90
  tiles 89–90
vapour barrier 82, 102
variations 26
VAT 21, 26–7
VAT receipt 27
VAT registration 27
ventilation 44, 55, 95, 100
verge tiles 91
vertical damp proof 65–7
vitrified clay pipes 43

wall plate 75–6, 78, 84, 87, 97,
  100
wall ties 63–4, 66
wall tiling 130–1
waste disposal 30, 34
waste disposal skips 34, 39
water bar 72
weedkiller 56
weep holes 68, 82
whacker plate 56
wheelbarrow 37, 60
windows 63–5, 68, 70, 82, 104,
  124
  aluminium 114
  boards 117–18
  frames 70, 76
  hardwood 114
  made to measure 71, 113
  upvc 114
Wiring Regulations 12